Industrial Explosion Prevention and Protection

Industrial Explosion Prevention and Protection

Frank T. Bodurtha
Engineering Department
E. I. du Pont de Nemours & Company
Incorporated

McGraw-Hill Book Company

New York
St. Louis
San Francisco
Auckland
Bogotá
Hamburg
New Delhi
London
Madrid
Mexico
Montreal
Johannesburg
Panama
São Paulo
Tokyo
Singapore
Sydney
Toronto

Library of Congress Cataloging in Publication Data

Bodurtha, Frank T

Industrial explosion prevention and protection.

Includes bibliographies and index.

1. Explosions—Safety measures. I. Title.

T55.3.E96B6 604'.7 79-24366

ISBN 0-07-006359-1

1 2 3 4 5 6 7 8 9 0 KPKP 8 9 8 7 6 5 4 3 2 1 0

The editors for this book were Jeremy Robinson, Elizabeth Richardson, and Carolyn Nagy, the designer was Mark E. Safran, and the production supervisor was Thomas G. Kowalczyk. It was set in Baskerville by ComCom.

Printed and bound by The Kingsport Press.

TO
SAFETY

Contents

Preface

On June 1, 1974, a massive explosion demolished a chemical plant in Flixborough, England, when about 50 Mg of cyclohexane was released to the atmosphere in approximately 40 s and then ignited. Twenty-eight lives were lost. This and other plant explosions highlight the urgent need for improved training of engineering students and engineers in the principles and practice of industrial explosion prevention and protection. This book has been written to help fill that need. Excellent *Bulletins* and *Reports of Investigations* of the Bureau of Mines, U.S. Department of Interior, and several books provide information and data on specific explosion topics. This text, on the other hand, supplies the fundamentals and other requirements to help the engineer analyze and solve the numerous explosion problems met in practice. This is particularly important in this era of increased public (and concomitant regulatory) attention to safety. Accordingly, the contents are intended for engineering students and for practicing engineers who have not received formal training in this important subject. The topics are treated in a quantitative manner insofar as possible. The material may be used in its entirety as a course or simply for solution of individual problems. The subject is presented in an orderly and logical manner to simplify the comprehension of this complex multidisciplinary subject and the solution to potential explosion hazards. The subject is broad, and not all aspects are covered; barricades are not included, for example. Also, risk analysis is not included, but this is a growing and useful tool for handling and decreasing potential explosion hazards. Moreover, some topics are not examined in detail, but numerous references are provided for additional information. The material will aid in recognition of potential explosion plus fire problems and in investigation of explosion incidents. Furthermore, it can help to allay fears when presumption of an explosion hazard is unjustified.

Known ignition sources should be eliminated to reduce the probability of an explosion. (Surprisingly, fires and explosions caused by welding and cutting still occur.) The presumed exclusion of ignition sources, however, should not be relied upon solely for safety except in unusual cases. Therefore, flammability limits are discussed first to emphasize the importance and methods of explosion prevention. (Inert gas is often used to exclude air and thereby prevent an explosion. In such cases, it is essential to guard against asphyxiation of personnel from inert gas that may leak out of equipment or be in vessels they enter.) When prevention cannot be practiced, or when additional safety is necessary due to the potential serious consequences of an explosion, protection measures against explosion effects must be implemented. Accordingly, explosion pressure and the methods of explosion protection are considered after a review of ignition sources. As plants become larger, the safe disposal of emergency discharges is a growing problem. Thus, atmospheric releases, including unconfined-vapor-cloud explosions, as occurred at Flixborough, are reviewed. Hazardous compounds, reactions, and operations are covered in Chap. 7. Accompanying the trend to large plants is the use of intricate conveying systems for dusts, with attendant explosion hazards. Therefore, dust explosions are covered at length in the final chapter.

A list of principal symbols is provided. This work is a multidisciplinary effort, and sometimes the same symbol has to be used for different meanings. As an example, V is used for potential differences in kilovolts in Chap. 3, but it is also the symbol for volume in other chapters. Thus, V is used as volume in the list Principal Symbols, and V is defined separately as potential difference in kilovolts where it is used in Chap. 3.

The International System of Units (SI) is used exclusively throughout this book. The author has used Standard E380 of the American Society for Testing and Materials as a guide. (This standard is also identified as the American National Standard Z210.1.) To help convert other units into the SI units, a list of conversion factors is given in Appendix A. *Note that moles are always gram moles.*

No attempt has been made to include a comprehensive list of flammability and explosion data. This information is published in several bulletins and standards.* For example, Fire Hazard Properties of Flammable Liquids, Gases and Volatile Solids, *NFPA* 325M, of the National Fire Protection Association contains flammability data for many chemicals. Nor are codes and standards or governmental laws and regulations included because of the increasing and changing nature of these requirements. Nevertheless, regulations of the Federal Occupational Safety and Health Administration (OSHA) and the individual states, of course, must be studied and complied with in the design and operation of process

*Extracts are given from the 1978 editions of the National Electrical Code and *NFPA* 68, Guide for Explosion Venting; the 1975 edition of *NFPA* 704, Standard System for the Identification of the Fire Hazard of Material; and the 1977 edition of Standard for Spray Application Using Flammable and Combustible Materials *NFPA* 33. These standards are all copyrighted by NFPA and have been extracted by permission.

plants. Moreover, it is essential to keep abreast of changes in these regulations to assure compliance.

The technical material herein is believed to be reliable and is intended for use by persons having technical skill. Since the conditions of use are beyond the control of the author, the author's employer, and the publisher, they assume no liability in connection with use of the information presented.

Frank T. Bodurtha

Principal Symbols

a = initial reactants, mol*
A = $SG^{2/3}/(SG - 1)^{1/3}$
A_v = explosion venting area, m^2
A' = horizontal cross-sectional area of vessel, m^2
A'' = surface area, m^2
c_p = specific heat at constant pressure, kJ/(kg) (K)
c_v = specific heat at constant volume, kJ/(kg) (K)
C_{st} = stoichiometric concentration, % v/v
D = stack exit diameter, mm
D_h = diameter of hole, mm
E = energy, MJ
E_a = Arrhenius activation energy, kJ/mol
Fr = Froude number, dimensionless
g = acceleration due to gravity, 9.806 m/s^2
h_s = stack height, m
H = plume rise, m
ΔH = heat release, MJ/kg or kJ/mol (negative ΔH = heat evolved; positive ΔH = heat absorbed)
ΔH_c = heat of combustion, MJ/kg or kJ/mol
ΔH_d = heat of decomposition, MJ/kg or kJ/mol
ΔH_f = heat of formation, kJ/mol
i_r = normal reflected pressure impulse, kPa·ms
i_s = incident (side-on) pressure impulse, kPa·ms
k = rate constant, s^{-1}
K = resistance coefficient or velocity head loss
K_G = constant for gases, bar·m/s
K_{St} = constant for dusts, bar·m/s
L = lower flammability limit, % v/v
L_t = lower flammability limit at t°C, % v/v
L' = lower flammability limit, mg/L air
MEC = minimum explosible concentration for dust, mg/L air
MOC = minimum oxygen for combustion, % v/v

*In this book moles are always gram moles.

MOC_P = minimum oxygen for combustion at P MPa, % v/v
MW = molecular weight, g/mol
n_f = final number of moles of gas in burned or decomposed mixture
n_i = number of moles of gas in initial mixture
p_b = burst pressure of rupture disk, kPa gage
p_i = initial pressure, kPa gage
p_m = maximum explosion pressure, MPa gage
p_r = peak positive normal reflected pressure, kPa gage
p_{so} = peak positive incident (side-on) pressure, kPa gage
p_v = maximum vented explosion pressure, kPa gage
P = absolute pressure, kPa or MPa
P_b = burst pressure of vessel, MPa abs
P_i = initial pressure, kPa abs
P_m = maximum explosion pressure, kPa abs
P_s = pressure in surrounding air, kPa or MPa abs
ΔP = differential pressure between two points,* kPa
Q = emission rate of gas or vapor, m^3/s at 25°C
Q_D = dust flow, kg/h
r = rate of explosion pressure rise dP/dt in closed vessel, MPa/s
r_{av} = average rate of explosion pressure rise $(dP/dt)_{av}$ in closed vessel, MPa/s
r_m = maximum rate of explosion pressure rise $(dP/dt)_m$ in closed vessel, MPa/s
R = distance from center of blast, m
R_u = universal gas constant, 0.008314 kJ/(mol) (K)
S_u = burning velocity, m/s
SG = specific gravity of emission (air = 1)
t = temperature, °C, or time, s
t_a = temperature of surroundings, °C
t_0 = duration of positive phase of pressure, ms
t_r = temperature of vessel contents, °C
T = thermodynamic temperature, K
T_d = adiabatic decomposition temperature, K (temperature resulting from adiabatic decomposition)
T_f = final temperature, K
T_i = initial temperature, K
T_s = stack-gas temperature, K
T_A = temperature of surroundings, K
T_R = temperature of vessel contents, K
u = mean wind speed, m/s
U = upper flammability limit, % v/v
U_t = upper flammability limit at t°C, % v/v
U_P = upper flammability limit in air at P MPa, % v/v
U' = heat-transfer coefficient, kJ/(s) (m^2) (K)
v_s = stack-gas exit velocity, m/s
$v_{s,crit}$ = critical stack-gas exit velocity, m/s
v_t = terminal velocity of particle, m/s

*P at outlet of a pipe can be greater than 101.325 kPa abs if velocity is sonic.

V = volume, m^3
V_f = volume of fuel $>L$, m^3 at 25°C
w = rate of flow of gas, kg/s
W = weight of detonable material, kg
W_c = weight of fuel available as explosive source, kg
W_{TNT} = TNT equivalent weight, kg
x = downwind distance, m
x_{m_1} = downwind distance from fictitious source to maximum height for specified L, m
x_{m_2} = downwind distance from true source to maximum height for specified L, m
x_L = downwind distance to L at ground, m
y = crosswind distance, m
Y = net expansion factor for compressible flow through orifices, nozzles, or pipe
z = height above ground level, m
Z = scaled distance, $m/kg^{1/3}$; Arrhenius preexponential factor, s^{-1}
α = "yield" of vapor-cloud explosions, dimensionless
κ = ratio of specific heats c_p/c_v
λ = depth of dust layer, mm
ρ = density, kg/m^3
σ_y = standard deviation in crosswind direction of plume concentration distribution, m
σ_z = standard deviation in vertical direction of plume concentration distribution, m
χ = atmospheric concentration, % v/v
χ_m = maximum atmospheric concentration, % v/v
χ_s = stack-gas concentration, % v/v
χ_D = dust concentration, mg/L

1

Introduction

The population of the United States increased from approximately 5 million in 1800 to 76 million in 1900. Explosives paved the way for the previously unimaginable industrial progress of that century. They helped in channeling numerous canals, e.g., the Erie, Illinois, and Michigan, to connect major waterways of the growing country.[1]* Later, explosives helped to bind the country together with countless railroads when this form of transportation proved to be preferable[1]; explosives open mines and oil wells. In these and numerous other ways, they have been a boon. Explosives have declined steadily in industrial prominence in the twentieth century, but accompanying the phenomenal industrial progress of this century, accidental explosions have been of great concern. Explosions account for the majority of property losses in the chemical and allied industries. Although no firm figures are available, annual property loss from industrial explosions in the United States is estimated to be over $150 million (1979 dollars). Losses from business interruption are comparable. Although safety principles and practices are in use that prevent this value from being even higher, added attention to the technology that exists and new knowledge are needed to reduce explosions and their effects. Methods to accomplish these goals are examined in the ensuing sections of this introductory chapter. Implementation of government regulations on the environment have increased fire and explosion risks in industrial plants, and this subject is also discussed.

1-1 The Role of Personnel in Explosion Prevention and Protection

The Role of Management

A persistent commitment to fire and explosion control by management is essential to limit losses. Safety performance is an excellent

*Numbered references are given at the end of the chapter.

yardstick of overall performance; safe operations are well-managed operations.

High production goals sometimes can result in explosions if shortcuts are taken; the resultant downtime is the opposite to the desired production goal. Also, the spacing and size of new process plants require close surveillance by management. Ample separation of various plant units is the best way to limit potential fire and explosion losses. The larger the plant equipment the greater the property and the business-interruption losses are likely to be. In addition, the odds increase for a fire or explosion with large equipment, considering the likelihood, for example, of a greater emission of flammable materials that can be ignited if equipment fails. Moreover, proper training of operating and maintenance personnel is essential to control fires and explosions satisfactorily. New personnel must be properly initiated into the intracacies of operating and maintaining a large process plant. Periodic retraining may be needed in some cases. Finally, tests and research should be supported when safety questions remain.

The Role of the Plant Engineer

From the standpoint of design and maintenance of production facilities, the plant engineer plays a key role in fire and explosion control.

Standards. Numerous national standards exist to aid in preventing fires and explosions. The National Fire Protection Association publishes annually a set of volumes of National Fire Codes on various aspects of fire and explosion control. Some codes have been adopted by the Occupational Safety and Health Administration (OSHA) as consensus standards. Also, the Factory Mutual Engineering Corporation provides a large number of Loss Prevention Data Sheets on explosions and explosion-related topics. In addition, the regulations of OSHA plus states and municipalities have to be followed. Furthermore, it is prudent for local plants to develop their own standards for solutions to recurring problems.

Design. Designs should be *fail-safe,;* i.e., upon any type of failure of its control system, the unit is automatically shut down or other actions are taken which make it impossible for the operation to become unsafe.[2] Explosion potential should be examined and remedies devised in the planning stage of a project. It cannot be merely presumed that safety will follow from the adopted design. When the safe process conditions are determined, it is highly important to monitor to be sure that these safety conditions are met while the plant is operating. Thus, it is often necessary, for example, to provide combustible-gas analyzers to aid in assuring the absence of flammable mixtures in plant equipment.

Fault-tree analysis may be used to uncover possibly hitherto unrecognized risks and thus reduce hazards.[3–11]

Maintenance. Dependable equipment is an ally in explosion control; preventive maintenance should be used in the campaign against explosions and resultant loss of production facilities. Moreover, when maintenance is performed, the proper parts, made from the right material, must be installed in the correct manner. The repair procedure itself should not impose fire and explosion risks. Thus, for example, local plant permits should be required for welding and cutting to lessen the risk of ignition.

Above all, the effect of all changes in design and materials of construction on safety should be examined closely; the consequence of the change on the entire facility should be evaluated. Seemingly minor alterations may impose unexpected and unacceptable explosion risks. The failure of a temporary 500-mm-diameter bypass dogleg pipe that was installed to replace a leaking reactor precipitated the Flixborough, England, explosion disaster. Accordingly, the monetary value of the change is not necessarily the primary criterion of the possible hazard produced.

Individual Responsibility

Individual responsibility is the cornerstone on which safety rests. Management must provide the means for explosion control, but industrial accidents often happen because established procedures are violated. Operating and maintenance personnel must be receptive to training and perform their assigned duties as prescribed. To do otherwise is to risk property loss and jeopardize the well-being of their fellow employees.

1-2 Risks from Pollution Control

Increased explosion risks resulting from the advent of stringent pollution-control regulations have been emphasized by LeVine[12] and Bodurtha.[13] From the air-quality standpoint, control of particulate matter, nitrogen oxides, and hydrocarbons sometimes cause explosion hazards. Proper recognition of the hazard of the total system and not just the pollution-control device itself is essential to safety.[13] As an example, the chances of generating flammable gases in an upstream production unit and then developing consequent explosion hazards from those gases in a combustion-type abater would need to be assessed.

Many devices are available for cleanup of particulate matter. Because of water-pollution control requirements that limit discharge of waterborne pollutants, there is a growing trend toward dry collection devices, such as bag houses and electrostatic precipitators, in place of scrubbers. Where combustible dusts are involved, explosions or fires have occurred in such installations. Also, caution is necessary when operating electrostatic precipitators on boiler-furnaces. When overfiring coal with oil, for example, fuel-rich conditions may develop; then fires can occur in the electrostatic precipitator where air is added.

There are several ways to achieve low emission of nitrogen oxides (NO_x) from boiler-furnaces.[14] They include low excess-air firing, multistage air admission, flue-gas recirculation, reduced secondary-air temperature, and, on new units, new designs such as wider spacing of burners.[13,14] These methods may impose unacceptable explosion risks if adequate procedures to prevent unsafe operations are not adopted.[13,14] The safety of low-NO_x emission regulations needs to be weighed by the environmental regulators.

Increased hydrocarbon-abatement requirements for control of ozone in the ambient air will often require combustion in one form or another.[13] Here, again, the safety of the combustion method used must be examined and the proper design and operating procedures developed to provide safety.

1-3 Terminology

There is some confusion over use of the terms vapor, gas, flammable, flammability, explosion, explosive, and explosible limits. In this text a *gas* means a substance that exists only in the gaseous state at 0°C and 101.325 kPa (1 standard atm = 101.325 kPa). *Vapor,* on the other hand, emanates from a substance that is a liquid at standard conditions. (The National Fire Protection Association defines a *flammable liquid,* in part, as one having a vapor pressure not exceeding 276 kPa abs at 37.8°C.) The terms *flammable, flammability, explosion,* and *explosive limits* are often used interchangeably; *explosible* is used in Western Europe. In conformity with current practice, however, such limits are referred to here as *flammability limits* for gases and vapors; dust is *explosible.*

An *explosion* is the result, not the cause, of rapid expansion of gases. It may occur from a physical or mechanical change, as in a boiler explosion, or by a chemical reaction. A *deflagration* is a reaction which propagates to the unreacted material at a speed that is less than the speed of sound in the unreacted substance. (Unless defined otherwise, an explosion is a deflagration in this text.) A *detonation* is an exothermic reaction that proceeds to the unreacted substance at a speed greater than the speed of sound. It is accompanied by a shock wave in the material and inordinately high pressure.

References

1. Wilkinson, N. B., *Explosives in History,* The Hagley Museum, Wilmington, Del., 1966.

2. Fawcett, H. H., and W. S. Wood, *Safety and Accident Prevention in Chemical Operations,* Interscience Wiley, New York, 1965.

3. Anon., "Risk Analysis Makes Chemical Plants Safer," *Chem. Eng. News,* vol. 56, no. 40, p. 8, Oct. 2, 1978.

4. Barlow, R. E., and P. Chatterjee, "Introduction to Fault Tree Analysis," *USNTIS AD Rep. AD774072,* December 1973.

5. Brown, D. B., *Systems Analysis and Design for Safety,* Prentice-Hall, Englewood Cliffs, N.J., 1976.

6. Browning, R. L., "Use a Fault Tree to Check Safeguards," *Chem. Eng. Prog. 12th Loss Prev. Symp., Atlanta, 1979,* pp. 20–26.

7. Katz, M. J., "Hazard and Risk Evaluation," *Chem. Eng. Prog. 10th Loss Prev. Symp., Kansas City, 1976,* pp. 127–134,

8. Kolodner, H. J., "Use a Fault Tree Approach," *Hydrocarbon Process,* vol. 56, no. 9, pp. 303, 304, 306, 308, September 1977.

9. Lambert, H. E., "Fault Tree for Locating Sensors in Process Systems," *Chem. Eng. Prog.,* vol. 73, no. 8, pp. 81–85, August 1977.

10. Menzies, R. M., and R. Strong, "Some Methods of Loss Prevention," *Chem. Eng. (Lond.),* no. 342, pp. 151–155, March 1979.

11. Powers, G. J., and F. C. Tompkins, "Fault Tree Synthesis for Chemical Processes," *AIChE J.,* vol. 20, no. 2, pp. 376–387, March 1974.

12. LeVine, R. Y., "Impact of Environmental Regulations on Loss Prevention," *Chem. Eng. Prog. 6th Loss Prev. Symp., San Francisco, 1972,* pp. 130–134.

13. Bodurtha, F. T., "Explosion Hazards in Pollution Control," *Chem. Eng. Prog. 10th Loss Prev. Symp., Kansas City, 1976,* pp. 88–90.

14. National Fire Protection Association, Standard for Prevention of Furnace Explosions in Pulverized Coal-Fired Multiple Burner Boiler-Furnaces, *NFPA* 85E, Boston, 1978.*

*Publications of the NFPA are updated on an irregular basis; dates are given when reference is to a specific version.

2

Flammability Limits

Where feasible, it is normally best to operate processes outside the range of flammability or with depletion of oxygen, i.e., to prevent explosions and fires. This can be accomplished by regulating temperature, airflow, and other process variables.

2-1 Flash Point

The *flash point* of a liquid is the minimum temperature at which it gives off sufficient vapor to form an ignitable mixture with air near the surface of the liquid or within the vessel used. (An *ignitable mixture* is a mixture within the range of flammability that is capable of the propagation of flame away from the source of ignition when ignited.[1]) Flash points are measured in closed and open cups by the methods indicated in Table 2-1.

In the tests, an open flame is used as an igniter. A discussion on the determination of flash point of chemicals by closed-cup methods is given in ASTM E502.

A fire will not necessarily develop at the flash point. The fire point is the lowest temperature at which liquid in an open container will give off enough vapor to continue to burn when once ignited. It is usually slightly above the open-cup flash point.[2] Open-cup flash points are higher than closed-cup flash points and are applicable, for example, to conditions above flammable liquids in open vessels and in spills.

The flash point of a flammable liquid is a fundamental and important property relative to fire and explosion hazards. Unfortunately it is often confused with *ignition temperature,* which is the temperature required to ignite the substance. Thus, the flash point is the temperature at which the vapors over a flammable liquid can be ignited. The closed-cup flash point is the temperature at which the equilibrium concentration of a vapor over

TABLE 2-1 Standard Methods for Determination of Flash Points

Method	Use	ASTM designation
Tag closed tester	For testing by Tag closed tester of liquids with a kinematic viscosity of below 5.5×10^{-6} m²/s at 40°C or below 9.5×10^{-6} m²/s at 25°C and a flash point below 93°C except cutback asphalts, liquids which tend to form a surface film under test conditions, and materials which contain suspended solids	D56
Cleveland open cup	For testing all petroleum products except fuel oils and those having an open-cup flash point below 79°C	D92
Pensky-Martens closed tester	For testing by Pensky-Martens closed-cup tester of fuel oils, lube oils, suspension of solids, liquids that tend to form a surface film under test conditions, and other liquids	D93
Tag open cup	For testing by Tag open-cup apparatus of liquids having flash points between −17.8 and 168°C	D1310
Setaflash closed tester	For testing by Setaflash closed tester of paints, enamels, lacquers, varnishes, and related products and their components having flash points between 0 and 110°C having a viscosity lower than 150 St at 25°C (the method can be used to determine whether a material will or will not flash at a specified temperature or to determine the finite temperature at which a material will flash)	D3278

a flammable liquid is equal to the lower flammability limit L of the vapor. As an example, the closed-cup flash point of toluene is 4.44°C, at which temperature its saturated vapor pressure is 1.24 kPa. At atmospheric pressure (101.325 kPa) this is equivalent to 1.22% volume/volume (v/v), compared with the L for toluene of 1.2% v/v.*

Closed-cup flash points of hydrocarbons can be estimated from their boiling points by the formula[2]

$$t_F = 0.683 t_B - 71.7 \tag{2-1}$$

where t_F = closed-cup flash point, °C
t_B = initial boiling point, °C

*Unless noted otherwise, lower flammability limits L in air are from Fire Hazard Properties of Flammable Liquids, Gases, Volatile Solids 1969, *NFPA* 325M, at normal atmospheric temperature and pressure. These values are essentially the same as in the 1977 edition.

If the temperature of a liquid is below its flash point, flammable concentrations of vapor cannot exist. Nevertheless, if mists form, conditions may become flammable below the flash point. Mists can develop by cooling or by mechanical means, such as spraying. Also, foams may be flammable even though they are at a temperature below the flash point. The *L* of fine mists plus accompanying vapor is about 48 mg of mist per liter of air at 0°C and 101.325 kPa.[3,4] At this concentration, mist is very dense and a 100-W bulb is visible in it only for a matter of inches.[5] Also, if a spark persists long enough, it may produce enough vaporization from a high-flash-point liquid to cause ignition.

Effect of Pressure

An increase in pressure raises the flash point, and a decrease in pressure lowers the flash point. Pressure effects on flash point are illustrated in Table 2-2, using toluene as an example.

Flammable liquids in mountainous areas have lower flash points than at sea-level locations. Denver, Colorado, at an elevation of 1.625 km, has an average atmospheric pressure of 83.6 kPa. In Denver, the flash point of toluene is 1°C.

Organic Aqueous Solutions

Organic materials are often in solution with water. There are few data on flash points of organic aqueous solutions. Raoult's law can be used to estimate the closed-cup flash points of these solutions, as discussed by Johnston.[6] The *L* of methyl alcohol is 6.70% v/v and for 75 wt % methyl alcohol in water the vapor space must contain (0.067)(101.325) = 6.79 kPa methyl alcohol at the closed-cup flash point. For 75 wt % methyl alcohol, the mole fraction of methyl alcohol in solution is 0.628. Thus, by Raoult's law the solution must be at a temperature high enough for the methyl alcohol by itself to have a saturated vapor pressure of 6.79/0.628 = 10.8 kPa. This closed-cup flash point is 17.0°C, whereas the closed-cup flash point of methyl alcohol is 11.1°C.

For high flash points, the effect of water vapor has to be considered.

TABLE 2-2 Effect of Pressure on Flash Point

Total pressure, kPa	Saturated vapor pressure of toluene at *L**, kPa	Calculated closed-cup flash point of toluene, °C
101.325	1.22	4.2
200	2.43	16.0
75	0.912	−0.5

**L* = 1.2% v/v, and no significant differences will exist at the specified pressures.

For example, at 83°C, the saturated vapor pressure of water is 52.7% v/v. With this water-vapor content, methyl alcohol and many organic materials are not flammable and they would have no closed-cup flash points. As will be discussed in subsequent sections, however, the materials may burn when air is added and therefore could have open-cup flash points.

The method for calculation of flash points described above applies to solutions that obey Raoult's law. For all mixtures, Raoult's law holds for a component when that component approaches 100 percent in solution. The ideality of a solution should be checked before estimating flash points by the previous method. Nevertheless, if the mole fraction of the organic material in solution is 0.8 or more, Raoult's law holds with an error of 7 percent or less except in extremely unusual cases.[7]

Also, a very small concentration of a volatile substance in an otherwise high-flash-point compound can yield a mixture with a low flash point. As an example, consider 2 wt % acetaldehyde in ethylene glycol, which has a high flash point of 111.1°C. (The lower flammability limit of acetaldehyde in air is 4.0% v/v.) Assume ideality. The mole percentage of acetaldehyde in solution is 2.80 percent. Thus, by Raoult's law at the flash point of the mixture (the vapor pressure of ethylene glycol is negligible)

$$0.028P = (0.040)(101.325)$$

where P is the vapor pressure of acetaldehyde in kilopascals and 101.325 kPa is standard atmospheric pressure.

The resulting vapor pressure of acetaldehyde is 145 kPa, corresponding to a closed-cup flash point of 29°C for the mixture.

2-2 Lower Flammability Limit *L*

The lower flammability limit is the minimum concentration of vapor or gas in air or oxygen below which propagation of flame does not occur on contact with a source of ignition.[1] Burgess and Wheeler[8] have shown that the heat generated by a mixture at the lower limit is substantially constant for many combustible-air mixtures. Spakowski[9] obtained the value of 4.354×10^3 for the product of the lower limit (volume percent) and the net heat of combustion in kilojoules per mole.*

The important criterion for lower limits is the ability of the mixture to propagate flame away from the source of ignition. Explosion pressure developed in small test apparatus is not an exact criterion of flammability; explosion of kernels of fuel at the ignition point may yield some pressure, which does not necessarily signify propagation of flame. Since more fuel may be needed for propagation at L, L's from explosion pressure are on

*In this book moles are always gram moles.

the safe side. Several conditions may affect determination of L's,[10] including the ignition source and the diameter and length of the test vessel. Upward flame propagation usually gives smaller L's than downward propagation. ASTM E681 is a test method for L's and U's.

Lower limits are usually expressed in volume percent, and as molecular weight increases, L's decrease. Calculated L's may be adequate when process compositions are well below the L. On a mass basis, L's for hydrocarbons are fairly uniform at about 45 mg/L air at 0°C and 101.325 kPa. [Alcohols and other oxygen-containing compounds have higher values. Ethyl alcohol (C_2H_5OH) = 70 mg C_2H_5OH per liter of air. Hydrogen has a much lower value, and other low-molecular-weight materials have slightly lower values.[11]] This near uniformity is illustrated in Table 2-3.

Consequently, ventilation rates to reduce concentrations of equal masses of different hydrocarbons to a specified percent of the lower limits are about the same.

Jones[12] first determined that lower limits for organics in air are about 55 percent of their stoichiometric concentration in air C_{st}. The combustion of organic compounds containing only carbon, hydrogen, and oxygen and a method for determination of C_{st} and estimation of L are given by the following equations:

$$C_nH_xO_y + \left(n + \frac{x}{4} - \frac{y}{2}\right)O_2 \rightarrow nCO_2 + \frac{x}{2}H_2O \tag{2-2}$$

$$\text{Air} = 4.77\left(n + \frac{x}{4} - \frac{y}{2}\right) \text{ mol/mol fuel} \tag{2-3}$$

$$\text{Nitrogen from air} = 3.77\left(n + \frac{x}{4} - \frac{y}{2}\right) \text{ mol/mol fuel} \tag{2-4}$$

$$C_{st} = \frac{100\% \text{ v/v}}{4.77n + 1.19x - 2.38y + 1} \tag{2-5}$$

$$L \approx \frac{55\% \text{ v/v}}{4.77n + 1.19x - 2.38y + 1} \tag{2-6}$$

For toluene ($C_6H_5CH_3$) $n = 7$, $x = 8$, and $y = 0$, and so

$$C_{st} = 2.28\% \text{ v/v}$$
$$L \approx 1.25\% \text{ v/v}$$
$$\approx 52 \text{ mg/L air (0°C and 101.325 kPa)}$$

TABLE 2-3 Uniformity of Lower Limits on Mass Basis

	Molecular Weight	L, % v/v	L, mg/L air*
Ethane	30	3.0	41
Propane	44	2.2	44
Hexane	86	1.1	43
Toluene	92	1.2	50

*At 0°C and 101.325 kPa.

2-3 Le Châtelier's Rule

Le Châtelier's rule[13] can be used to calculate the composite L of flammable mixtures:

$$\text{Composite } L = \frac{100\% \text{ v/v}}{c_1/L_1 + c_2/L_2 + c_3/L_3 + \cdots + c_n/L_n} \tag{2-7}$$

where $c_1, c_2, c_3, \ldots, c_n$ = percentages of volume of total combustibles ($c_1 + c_2 + c_3 + \cdots + c_n = 100$)

$L_1, L_2, L_3, \ldots, L_n$ = lower flammability limit of each combustible, % v/v

An example of an application of Le Châtelier's rule follows.

Example

A combustible-air mixture has the following composition:

	% v/v	
Hexane	0.8	$L_1 = 1.1\%$ v/v
Methane	2.0	$L_2 = 5.0$
Ethylene	0.5	$L_3 = 2.7$
Total combustible	3.3	
Air	96.7	

Therefore,

$$\text{Hexane} = \frac{0.8}{3.3}(100) = 24.2\% \text{ of total combustibles} = c_1$$

$$\text{Methane} = \frac{2.0}{3.3}(100) = 60.6 \text{ of total combustibles} = c_2$$

$$\text{Ethylene} = \frac{0.5}{3.3}(100) = 15.2 \text{ of total combustibles} = c_3$$

$$\text{Composite } L = \frac{100}{24.2/1.1 + 60.6/5.0 + 15.2/2.7}$$

$$= 2.5\% \text{ v/v vs } 3.3\% \text{ v/v total combustible}$$

The mixture is flammable even though the concentration of each constituent is less than its lower limit.

Among other mixtures, the rule works well with lower limits of solvent mixtures containing methyl ethyl ketone and tetrahydrofuran.[14] Nevertheless, deviations have been measured. For example, the rule does not hold particularly well for L's of some mixtures of hydrogen sulfide and carbon disulfide.[15] Thus, Le Châtelier's rule should be applied with discretion, particularly for chemically dissimilar compounds.

Composite upper flammability limits can be estimated in a similar manner.

2-4 Upper Flammability Limit U

The upper limit U is the maximum concentration of vapor or gas in air above which propagation of flame does not occur on contact with a source of ignition. The range of concentrations between the lower and upper limits is known as the *range of flammability.* Explosion prevention can be practiced, for example, by operating outside this range in processes using air. Nevertheless, flammable and dangerous conditions may develop if combustibles are lost when ostensibly operating above the upper limit. For many compounds, the upper limits are about 3.5 times the stoichiometric concentration in air.

Cool flames may occur above U. Cool flames result from a relatively slow, scarcely visible reaction. They are due to the decomposition of hydroperoxides, formed by low-temperature oxidation. Cool flames are not normally associated with brief ignition sources, such as sparks.[11]

2-5 Minimum Oxygen for Combustion (MOC)

If the oxygen content of a combustible mixture is decreased sufficiently, flame will not propagate. Therefore, explosion prevention can also be accomplished by adequate depletion of oxygen *whatever the concentration of the combustible.* At L, oxygen is in excess for combustion. Also, as noted earlier, the products of lower limits and net heats of combustion are substantially constant for many combustible-air mixtures. Since dry air is 79.05% v/v nitrogen, nitrogen and air have similar thermal conductivities, heat capacities, and molecular weights. Consequently, if nitrogen is substituted for some air, the same amount of heat will be generated at L until stoichiometric conditions are reached. Any further decrease in oxygen will result in less heat generation, which will be insufficient for flame propagation.

TABLE 2-4 Minimum Oxygen for Combustion (MOC)* (Data from Ref. 16)

	N_2–Air, % v/v O_2†	CO_2–Air, % v/v O_2†
Acetone	13.5	15.5
Benzene	11	14
Butadiene	10	13
Butane	12	14.5
Carbon disulfide	5	8
Carbon monoxide	5.5	6
Diethyl ether	10.5	13
Ethyl alcohol	10.5	13
Ethylene	10	11.5
Hydrogen	5	6
Hydrogen sulfide	7.5	11.5
Isobutane	12	15
Methane	12	14.5
Methyl alcohol	10	13.5
Propane	11.5	14
Propylene	11.5	14

*Safety factors for industrial operations are required.

†% v/v O_2 is in mixtures of the combustible + inert gas + air. Values are for normal room temperature and 101.325 kPa.

Minimum oxygen for combustion (MOC) for selected compounds are tabulated in Table 2-4.

In general, organic combustible compounds will not propagate flame if O_2 in mixtures of the organic, inert gas, and air is below about 10.5 and 13% v/v with nitrogen and carbon dioxide, respectively, as the inert gases. Cool flames are not likely to occur below the MOC.

Minimum oxygen for combustion with nitrogen as the inert gas can be calculated with good (and generally sufficient) accuracy from the oxygen required for complete combustion at the lower limit; any further decrease in oxygen will prevent ignition. For example, with carbon monoxide (CO),

$$CO + 0.5O_2 \rightarrow CO_2$$
$$L = 12.5\% \text{ v/v CO in air}$$
$$\text{MOC} = (12.5)(0.5) = 6.25\% \text{ v/v } O_2 \text{ (5.5\% in Table 2-4)}$$

propane (C_3H_8)

$$C_3H_8 + 5O_2 \rightarrow 3CO_2 + 4H_2O$$
$$L = 2.2\% \text{ v/v } C_3H_8 \text{ in air}$$
$$\text{MOC} = (2.2)(5) = 11.0\% \text{ v/v } O_2 \text{ (11.5\% in Table 2-4)}$$

and methyl alcohol (CH_3OH)

$$CH_3OH + 1.5O_2 \rightarrow CO_2 + 2H_2O$$
$$L = 6.7\% \text{ v/v } CH_3OH \text{ in air}$$
$$\text{MOC} = (6.7)(1.5) = 10.0\% \text{ v/v } O_2 \text{ (10\% in Table 2-4)}$$

Water vapor is an acceptable inert gas provided the temperature in the gas space is high enough. Also, water may contaminate some processes. There are few data on the MOC with water vapor as the inert gas. Zabetakis[17] and Zabetakis and Jones[18] determined MOCs for H_2–water-vapor–air mixtures at 149°C and CS_2–water-vapor–air at 100°C, respectively. Their MOC values and the MOC for ethyl alcohol[3] are shown in Table 2-5. MOCs from Table 2-4 with nitrogen and carbon dioxide as inert gases are shown for comparison.

Because the heat capacity of water vapor is higher than nitrogen and only slightly less than carbon dioxide, MOCs with water vapor as the inert gas are between the corresponding values for nitrogen and carbon dioxide. Accordingly, oxygen content that is safe with nitrogen will also be safe with water vapor as the inert gas. Minimum temperatures for specified O_2 percentages with saturated water-vapor–air mixtures are indicated in Table 2-6.

Some halogen-containing compounds can also be used as inerting materials at relatively low concentrations. For example, mixtures of dimethyl ether–air–F-12 (CCl_2F_2) and ethyl mercaptan–air–F-12 are inerted with 17 and 13.3% v/v F-12, respectively.[19,20] The corresponding MOCs are 15.7 and 16.8% v/v. Caution must be exercised at higher pressure due to the possibility of combustion of some halohydrocarbons themselves. For instance, methyl bromide is only barely flammable at atmospheric pressure, but a violent explosion occurred with methyl bromide and air at elevated pressure.[21]

It is imperative to assure adequate depletion of oxygen if that is the method of explosion prevention. This can often be accomplished by operating at a slight positive pressure with inert gas automatically sup-

TABLE 2-5 Minimum Oxygen for Combustion with Water Vapor as Inert Gas

	N_2–air, % v/v O_2	H_2O–air, % v/v O_2	CO_2–air, % v/v O_2
Hydrogen	5	5.0 ± 0.2 (149°C)	6
Carbon disulfide	5	7.6 (100°C)	8
Ethyl alcohol	10.5	12.3 (100°C)	13

TABLE 2-6 Minimum Temperatures for Specified O_2 with Saturated Water-Vapor–Air Mixtures at 101.325 kPa

Oxygen, % v/v	Minimum temperature, °C	Oxygen, % v/v	Minimum temperature, °C
5	92.6	10	82.8
6	90.9	11	80.5
7	89.1	12	77.9
8	87.1	13	75.0
9	85.1	14	71.8

plied at a set pressure by a pressure regulator. (See also the discussion of distillation columns in Sec. 7-3.) Storage tanks, for example, can collapse from negative pressure if the supply of inert gas fails; alarms should be provided to warn of loss of inert-gas pressure in the supply. Also, an oxygen analyzer may be used to regulate flow of inert gas to maintain safe oxygen; lag time in the sample and analyzer systems must be low. A flow switch with alarm should be used to signal loss of flow of sample gas to the analyzer. Alternatively, a portable oxygen analyzer can be used to establish the flow of inert gas required to give safe oxygen. Then a rotameter should be provided to measure inert-gas flow continuously. It should be equipped with an alarm to signal potentially dangerously low flow of inert gas as previously determined.

2-6 Flammability Diagrams

Flammability data can be shown on triangular or rectangular plots. They aid in interpreting the flammability consequences of changes in composition of process gases, particularly those developed by addition of air. Two

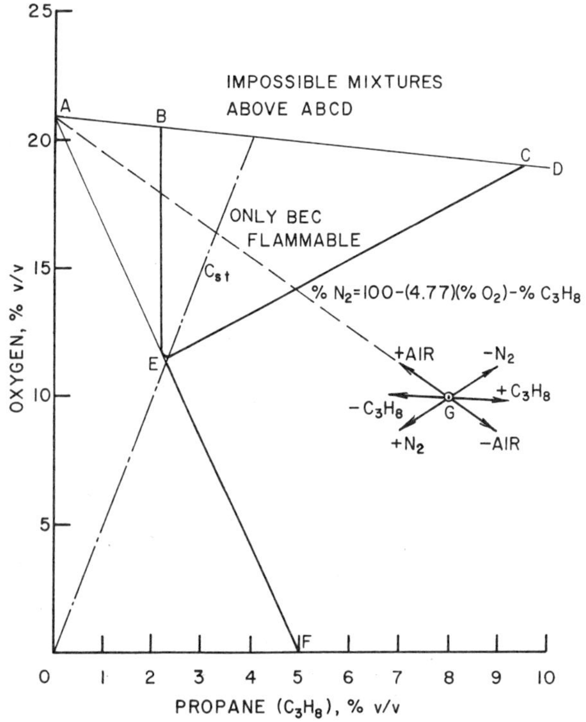

Fig. 2-1 Flammability diagram for mixtures of propane, air, and nitrogen at normal atmospheric temperature and 101 kPa abs (% N_2 = percentage of added N_2).

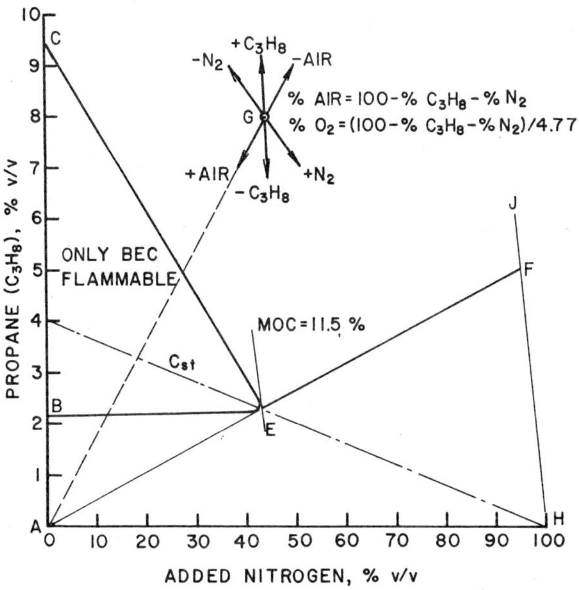

Fig. 2-2 Flammability diagram for mixtures of propane, air, and nitrogen at normal atmospheric temperature and 101 kPa abs (% added N_2 is in mixture of air + propane + added N_2).

types of rectangular plots for propane-air-nitrogen mixtures are shown in Figs. 2-1 and 2-2. (Flammability diagrams can be prepared for any compound with corresponding flammability limits for that compound.) Pertinent flammability data for propane are

$$L = 2.2\% \text{ v/v}$$
$$U = 9.5$$
$$C_{st} = 4.02$$
$$\text{MOC} = 11.5$$

In Fig. 2-1 all compositions of propane and air must fall along *ABCD*. In both Figs. 2-1 and 2-2, *B* is the lower limit *L* in air; *C* is the upper limit *U*. C_{st} is the line representing stoichiometric concentrations in propane-air-nitrogen mixtures. *Mixtures within BEC are the only flammable compositions.* The spokes at *G* illustrate how compositions change as propane, air, or nitrogen is added or subtracted; the positive spokes are aimed at 100% v/v of each constituent. Mixtures to the right of *FECD* are too rich to burn but can burn if air is added. Compositions at *G,* for example, are diluted along the straight line *GA* to infinite dilution at *A* (100% air) as air is added and in so doing pass through the flammable region. (Note that when the inert gas is water vapor, the line *GA* will bow toward *C* as air is added if water vapor condenses; additional air must replace condensed water vapor at constant pressure.) *F* is the maximum combustible content in a combustible-gas–inert-gas mixture that cannot burn when air is added in any amount. Also, F equals the tangent of angle $FAH \times 100$ in Fig. 2-2. For propane in nitrogen, $F = 5.0\%$ propane. Comparable values for several other combustible compounds are given in table 7 of Ref. 22.

Any mixture along or to the left of FEB cannot burn by itself or if air is added. The MOC is at E in Fig. 2-1. In Fig. 2-2 the MOC is determined by a line at the "nose" of the curve parallel to the zero-air line HFJ. Flammability diagrams for many compounds and additional explanations of them can be found in Refs. 3 and 10.

It is sometimes necessary to determine the flammability limits in air for a mixture of flammable gases in inert gas, initially with or without air. Flammability diagrams can be drawn to estimate the flammability limits of such mixtures. Consider a mixture containing 40 percent each of air and added nitrogen, plus 10 percent each of methane and hexane (G in Fig. 2-3).

Example

Lower and upper limits for each combustible in air are:

	L, % v/v	U, % v/v
Methane (CH_4)	5.0	15
Hexane (C_6H_{14})	1.1	7.5

In the combustibles, methane and hexane each equal 50% v/v

$$L = \frac{100}{50/5.0 + 50/1.1} = 1.8\% \text{ v/v methane + hexane in air}$$

$$U = \frac{100}{50/15 + 50/7.5} = 10\% \text{ v/v methane + hexane in air}$$

$$0.5CH_4 + O_2 \rightarrow 0.5CO_2 + H_2O$$
$$0.5C_6H_{14} + 4.75O_2 \rightarrow 3CO_2 + 3.5H_2O$$
$$(0.5CH_4 + 0.5C_6H_{14}) + 5.75O_2 \rightarrow 3.5CO_2 + 4.5H_2O$$

From Eq. (2-2), $n = 3.5$ and $x = 9$. (Total combustibles must equal 1 mol.) From Eq. (2-5), $C_{st} = 3.50\%$ v/v. MOC $= (1.8)(5.75) = 10\%$ v/v.

A flammability diagram for methane + hexane, air, and nitrogen is shown in Fig. 2-3. (At 20°C hexane does not have enough vapor pressure to exceed 15.8% v/v.) When air is added to G, the L and U of the total combustibles are 1.8 and 7.6% v/v, respectively, as shown at B' and C'. (Note that lower flammability limits do not decrease appreciably in oxygen-depleted atmospheres.) The L and U of the mixture in air are $(1.8\%)(100)/20 = 9.0$ percent and $(7.6\%)(100)/20 = 38$ percent, respectively, since total combustibles are 20 percent of G.

Special consideration is necessary for complicated mixtures containing several inert gases.

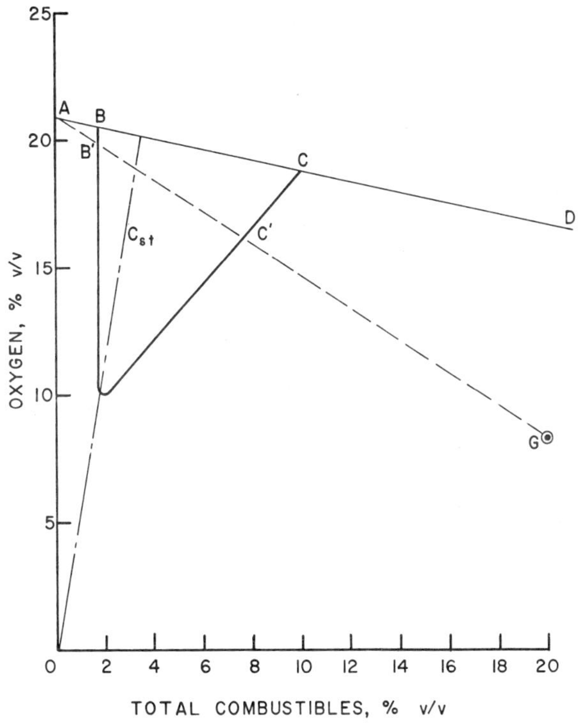

Fig. 2-3 Flammability diagram for mixtures of methane, hexane, air, and nitrogen at normal atmospheric temperature and 101 kPa abs (% methane = % hexane).

2-7 Environmental Effects on Flammability Limits

The flammability limits covered in the preceding sections are based on normal atmospheric temperature and pressure. Environmental factors have a pronounced effect on limits that must be considered in evaluating and decreasing process hazards.

Temperature

Lower flammability limits in air are decreased approximately 8 percent by a temperature increase of 100°C.[3,14,23,24] Upper flammability limits are increased 8 percent by a temperature increase of 100°C.[3,25] Thus, for t°C

$$L_t = L_{25°\mathrm{C}} - (0.8L_{25°\mathrm{C}} \times 10^{-3})(t - 25) \qquad (2\text{-}8)$$

$$U_t = U_{25°\mathrm{C}} + (0.8U_{25°\mathrm{C}} \times 10^{-3})(t - 25) \qquad (2\text{-}9)$$

Calculated and experimental L's for toluene and hexane at 200°C are shown in Table 2-7. (L's for toluene and hexane at 25°C are 1.24 and 1.26% v/v, respectively.) Temperature has a comparable effect on MOC.

Pressure

Pressure has only a slight effect on L. Lower limits are essentially constant down to about 5 kPa generally, below which pressure flame does not propagate. The effect of higher pressures on L and correspondingly on MOC is slight, too, as illustrated in Table 2-8.

The MOC with nitrogen for several saturated hydrocarbons from 0.1 to 13.8 MPa abs depends on the logarithm of the pressure such that

$$MOC_P \approx MOC - 1.5(\log P + 1) \qquad (2\text{-}10)$$

where P is absolute pressure in megapascals and MOC is at 0.101 MPa.[3]

Conversely, elevated pressure greatly increases upper flammability limits. For several saturated hydrocarbons from 0.1 to 20.7 MPa, U depends on the logarithm of the pressure:

$$U_P \approx U + 20.6(\log P + 1) \qquad (2\text{-}11)$$

where P is absolute pressure in megapascals and U is at 0.101 MPa.[26–29]

Gases with positive heats of formation can be decomposed explosively at high pressures in the absence of air with strong enough igniters and thus have no U's in these circumstances. Ethylene decomposes explosively at elevated pressure and acetylene at sea-level pressure in large-diameter piping.[30,31] Heats of formation for these materials are 52.3 and 227 kJ/mol, respectively.

Explosion prevention can be practiced by mixing decomposable gases with inert diluents. For instance, acetylene is rendered nonexplosive at a gage pressure of 100 kPa by 14.5% water vapor and by 8% v/v butane.[32]

TABLE 2-7 Calculated and Experimental Lower Flammability Limits in Air at 200°C[14,24]

	Toluene, % v/v	Hexane, % v/v
Calculated	1.07	1.08
Experimental	1.07	1.14

TABLE 2-8 Effect of Elevated Pressure on *L* and MOC* of Ethane in Air[26]

Pressure, MPa gage	L, % v/v	% decrease in L	MOC, % v/v	% decrease in MOC
0	2.85	. . .	11	
0.69	2.80	1.75		
1.72	2.70	5.20	9.3	15.5
3.45	2.55	10.5	8.9	19.1
5.17	2.40	15.8		
6.20	. . .	. . .	8.8	20.0
6.90	2.20	22.8		

*Nitrogen as inert gas.

Oxygen

Depletion of oxygen below the MOC prevents explosions, as already described.

Lower flammability limits in oxygen are about the same as in air, since oxygen in air is in excess for combustion at L. This similarity is evident in Table 2-9.

Upper limits increase markedly in oxygen-rich atmospheres, as is clear for pure O_2 in Table 2-9. Upper limits are related to oxygen concentration at 101 kPa, absolute, in the following manner:

$$U_{\%O_2} \approx U + 70(\log \%O_2 - 1.321) \qquad \text{where } \%O_2 = 20.95 \text{ to } 100 \tag{2-12}$$

where U is in air (20.95% v/v O_2). Percent O_2 v/v is in the fuel-free mixture. The U's, as in conventional use, are in the total mixture.

Other Oxidants

Chlorine. Flammable compounds are oxidized by other oxidants, such as chlorine. Flammability limits of hydrogen, methane, and ethane in chlorine at 25°C and 101.325 kPa are shown in Table 2-10.

Limits in chlorine are wider than in air and are comparable to the wide limits in oxygen. Chlorine is a common process gas, and its particular hazard is the relatively low autoignition temperatures of oxidizable substances in it. Thus, spontaneous ignition may occur at normal operating temperatures; in flammability tests, addition of ethane to chlorine at 200°C and 1.38 MPa gage caused spontaneous ignition.[34] The "explosion" temperature of 18% v/v H_2 in chlorine is 227°C,

TABLE 2-9 Flammability Limits in Air and Oxygen[10,33]

	L in Air, % v/v	L in O_2, % v/v	U in Air, % v/v	U in O_2, % v/v
Butane	1.9	1.8	8.5	49
Butene-1	1.6	1.8	9.3	58
Ethane	3.0	3.0	12.5	66
Ethylene	3.1	3.0	32	80
Hexane (95.5°C, 0.586 MPa gage)	1	1	17	65
Isopropyl ether	. . .	. . .	21	69
Methane	5.3	5.1	14	61
Propane	2.2	2.3	9.5	55
Vinyl chloride	4.0	4.0	22	70

TABLE 2-10 Flammability Limits in Chlorine[3]

	Lower limit, % v/v	Upper limit, % v/v
Hydrogen	4.1	89
Methane	5.6	70
Ethane	6.1	58

whereas the autoignition temperature of hydrogen in air is 400°C.[35]

Oxides of nitrogen. Flame can propagate in mixtures of oxides of nitrogen and oxidizable substances. Flammability limits of several compounds in nitrous oxide (N_2O) and nitric oxide (NO) are listed in Ref. 10. Flammability limits of several chlorinated hydrocarbons in nitrogen tetroxide*–nitrogen atmospheres are given in Ref. 36. Flammability limits of butane and ethylene dichloride in oxides of nitrogen at 101.325 kPa are shown in Table 2-11.

The minimum nitrogen tetroxide concentration for flame propagation with combustible gases in nitrogen tetroxide–nitrogen mixtures at 100°C and 101.325 kPa is about 25% v/v; butane requires 23% nitrogen tetroxide and ethylene dichloride 22% v/v.[36] As noted earlier, MOCs in mixtures of combustibles, nitrogen, and air are about 10.5% v/v. (There are only scanty data on minimum oxidant concentration for flame propagation with nitrous oxide and/or nitric oxide as the oxidants.) Moreover, the autoignition temperatures in nitrogen tetroxide are lower than in air or oxygen; for ethylene dichloride the autoignition temperatures in air, oxygen, and nitrogen tetroxide are 476, 470, and 282°C, respectively. The corresponding values for butane are 405, 283, and 270°C.[36]

TABLE 2-11 Flammability Limits in Oxides of Nitrogen[10,36]

	L, % v/v butane	*U*, % v/v butane	*L*, % v/v ethylene dichloride at 100°C	*U*, % v/v ethylene dichloride at 100°C
Nitrous oxide	2.5	20		
Nitric oxide	7.5	12.5		
Nitrogen tetroxide	. . .	. . .	13.7	48
Air	1.9	8.5	4.5	17.3
Oxygen	1.8	49	4.0	67.5

*Refers to equilibrium concentrations between nitrogen dioxide (NO_2) and nitrogen tetroxide (N_2O_4). At 100°C NO_2 = 95%.[36]

References

1. National Fire Protection Association, Fire Hazard Properties of Flammable Liquids, Gases, Volatile Solids, *NFPA* 325M, Boston, 1969.

2. Factory Mutual Engineering Corporation, *Handbook of Industrial Loss Prevention,* 2d ed., chap. 42, McGraw-Hill, New York, 1967.

3. Zabetakis, M. G., "Flammability Characteristics of Combustible Gases and Vapors," *U.S. Bur. Mines Bull.* 627 (*USNTIS* AD-701 576), 1965.

4. Burgoyne, J. H., "The Flammability of Mists and Sprays," *Inst. Chem. Eng.*

*Refers to equilibrium concentrations between nitrogen dioxide (NO_2) and nitrogen tetroxide (N_2O_4). At 100°C NO_2 = 95%.[36]

Symp. Ser. 15, *Proc. 2d Symp. Chem. Process Hazards Spec. Ref. Plant Des., 1963,* pp. 1–5.

5. Burgoyne, J. H., "Mist and Spray Explosions," *Chem. Eng. Prog.,* vol. 53, no. 3, pp. 121M–124M, March 1957.

6. Johnston, J. C., "Estimating Flash Points for Organic Aqueous Solutions," *Chem. Eng.,* vol. 81, no. 25, p. 122, Nov. 25, 1974.

7. Perry, J. H., *Chemical Engineers' Handbook,* 3d ed., sec. 8, McGraw-Hill, New York, 1950.

8. Burgess, M. J., and R. V. Wheeler, "Lower Limit of Inflammation of Mixtures of Paraffin Hydrocarbons with Air," *J. Chem. Soc.,* vol. 99, pp. 2013–2030, 1911.

9. Spakowski, A. E., Pressure Limit of Flame Propagation of Pure Hydrocarbon-Air Mixtures at Reduced Pressures, *Natl. Advis. Comm. Aeronaut. Res. Mem.* E52H15, 1952.

10. Coward, H. F., and G. W. Jones, "Limits of Flammability of Gases and Vapors," *U.S. Bur. Mines Bull.* 503 (*USNTIS* AD 701 575), 1952.

11. Burgoyne, J. H., "Principles of Explosion Prevention," *Chem. Process Eng. (Lond.),* vol. 42, no. 4, pp. 157–161, April, 1961.

12. Jones, G. W., "Inflammation Limits and Their Practical Application in Hazardous Industrial Operations," *Chem. Rev.,* vol. 22, no. 1, pp. 1–26, February 1938.

13. Le Châtelier, H., "Estimation of Firedamp by Flammability Limits," *Ann. Mines,* vol. 19, ser. 8, pp. 388–395, 1891.

14. Zabetakis, M. G., J. C. Cooper, and A. L. Furno, "Flammability in Air of Solvent Mixtures Containing Methyl Ethyl Ketone and Tetrahydrofuran," *U.S. Bur. Mines Rep. Invest.* 6048, 1962.

15. Miller, D. J., and C. W. Webb, Jr., "Lower Flammability Limits of Hydrogen Sulfide and Carbon Disulfide Mixtures," *J. Chem. Eng. Data,* vol. 12, no. 4, pp. 568–569, October 1967.

16. National Fire Protection Association, Standard on Explosion Prevention Systems, *NFPA* 69, Boston, 1973.

17. Zabetakis, M. G., Research on the Combustion and Explosion Hazards of Hydrogen–Water Vapor–Air Mixtures, Final Report, *AECU* 3327, 1956.

18. Zabetakis, M. G., and G. W. Jones, "Flammability of Carbon Disulfide in Mixtures of Air and Water Vapor," *Ind. Eng. Chem.,* vol. 45, no. 9, pp. 2079–2080, September 1953.

19. Jones, G. W., and F. E. Scott, "Inflammability of Dimethyl Ether–Dichlorodifluoromethane–Air Mixtures," *U.S. Bur. Mines Rep. Invest.* 4125, 1947.

20. Jones, G. W., M. G. Zabetakis, and G. S. Scott, "Elimination of Ethyl Mercaptan Vapor-Air Explosions in Stench Warning Systems," *U.S. Bur. Mines Rep. Invest.* 5090, 1954.

21. Hill, H. W., "Methyl Bromide–Air Explosion," *Chem. Eng. Prog.,* vol. 58, no. 8, pp. 46–49, August 1962.

22. American Gas Association, *Purging Principles and Practice,* New York, 1954.

23. Affens, W. A., "Flammability Properties of Hydrocarbon Fuels," *J. Chem. Eng. Data,* vol. 11, no. 2, pp. 197–202, April 1966.

24. Zabetakis, M. G., G. S. Scott, and G. W. Jones, "Limits of Flammability of Paraffin Hydrocarbons in Air," *Ind. Eng. Chem.*, vol. 43, no. 9, pp. 2120–2124, September 1951.

25. Rolingson, W. R., J. MacPherson, P. D. Montgomery, and B. L. Williams, "Effect of Temperature on the Upper Flammable Limit of Methane, Ammonia, and Air Mixtures," *J. Chem. Eng. Data,* vol. 5, no. 3, pp. 349–351, July, 1960.

26. Kennedy, R. E., I. Spolan, W. K. Mock, and G. S. Scott, "Effect of High Pressures on the Explosibility of Mixtures of Ethane, Air, and Carbon Dioxide and of Ethane, Air, and Nitrogen," *U.S. Bur. Mines Rep. Invest.* 4751, 1950.

27. Jones, G. W., R. E. Kennedy, and I. Spolan, "Effect of High Pressures on the Flammability of Natural Gas-Air-Nitrogen Mixtures," *U.S. Bur. Mines Rep. Invest.* 4457, 1949.

28. Kennedy, R. E., I. Spolan, and G. S. Scott, "Explosibility of Mixtures of Propane, Air, and Carbon Dioxide and of Propane, Air, and Nitrogen at Elevated Pressures," *U.S. Bur. Mines Rep. Invest.* 4812, 1951.

29. Zabetakis, M. G., "Fire and Explosion Hazards at Temperature and Pressure Extremes," *AIChE–Inst. Chem. Eng. Symp. Ser.* 2, *Chem. Eng. Extreme Cond. Proc. Symp., 1965,* pp. 99–104.

30. Scott, G. S., R. E. Kennedy, I. Spolan, and M. G. Zabetakis, "Flammability Characteristics of Ethylene," *U.S. Bur. Mines Rep. Invest.* 6659, 1965.

31. Sargent, H. B., "How to Design a Hazard-Free System to Handle Acetylene," *Chem. Eng.*, vol. 64, no. 2, pp. 250–254, February 1957.

32. Jones, G. W., R. F. Kennedy, I. Spolan, and W. J. Huff, "Effect of Pressure on the Explosibility of Acetylene-Water Vapor, Acetylene-Air, and Acetylene-Hydrocarbon Mixtures," *U.S. Bur. Mines Rep. Invest.* 3826, September 1945.

33. McGillivray, R. J., "Take These Steps to Sweetening Safety," *Hydrocarbon Process. Pet. Refiner,* vol. 42, no. 5, pp. 145–146, May 1963.

34. Bartkowiak, A., and M. G. Zabetakis, "Flammability Limits of Methane and Ethane in Chlorine at Ambient and Elevated Temperatures and Pressures," *U.S. Bur. Mines Rep. Invest.* 5610, 1960.

35. Kunin, T. I., and V. I. Serdyukov, "Chlorine and Hydrogen Explosion Temperatures and Limits in Hydrogen Chloride," *Zh. Obshch. Khim.*, vol. 16, no. 9, pp. 1421–1430, 1946.

36. Kuchta, J. M., H. L. Furno, A. Bartkowiak, and G. H. Martindill, "Effect of Pressure and Temperature on Flammability Limits of Chlorinated Hydrocarbons in Oxygen-Nitrogen and Nitrogen Tetroxide–Nitrogen Atmospheres," *J. Chem. Eng. Data,* vol. 13, no. 3, pp. 421–428, July 1968.

3

Ignition Sources

Flammable gases can be ignited by a host of sources. In the design and operation of processes, it is normally best not to base explosion safety on the often unwarranted presumption that ignition sources have been excluded. (In 318 natural-gas fires and explosions, 28 percent of the sources and forms of the heat of ignition were unknown.[1]) Nevertheless, all reasonable measures should be taken to eliminate them. Also, knowledge of the type and power of igniters is needed to help uncover the cause of explosions. A variety of ignition sources are discussed in the following sections.

3-1 Autoignition

The *autoignition temperature* (AIT) of a substance is the temperature at which vapors ignite spontaneously from the heat of the environment. It has also been defined as the spontaneous-ignition temperature. The minimum-autoignition, minimum-spontaneous-ignition, and self-ignition temperatures are the lowest temperature at which spontaneous ignition occurs. The autoignition temperature depends on many factors, namely, ignition delay, concentration of vapors, environmental effects (volume, pressure, and oxygen content), catalytic material, and flow conditions. Thus, a specified AIT should be interpreted as applying only to the experimental conditions employed in its determination. Moreover, in conducting tests to determine AITs for processes, all feasible steps should be taken to duplicate process conditions and equipment.

Ignition Delay

The time delay between the moment of exposure of a substance to high temperature and visible combustion is called *ignition delay.* Semenov[2] related these variables by the equation

$$\log \tau = \frac{52.55E}{T} + B \tag{3-1}$$

where τ = ignition delay, s
E = apparent activation energy, kJ/mol
B = constant

Also,

$$\log \frac{\tau_1}{\tau_2} = 52.55E \, \frac{T_2 - T_1}{T_2 \, T_1} \tag{3-2}$$

and

$$0.4343 \, \frac{d\tau}{\tau} = \frac{-52.55E}{T} \, \frac{dT}{T}$$

Thus,

$$\frac{d\tau}{\tau} = \frac{-121.0E}{T} \, \frac{dT}{T} \tag{3-3}$$

Although E is approximately 167.4 kJ/mol,[3,4] it varies somewhat between classes of compounds. Consequently, a 1 percent decrease in thermodynamic temperature generally results in a $(20.26 \times 10^3)/T$ percent increase in ignition delay.

In ASTM D2155, Standard Method of Test for Autoignition Temperature of Liquid Petroleum Products, the contents of a 200-mL Erlenmeyer flask are observed for the appearance of a flame for 5 min or until autoignition occurs. When ignition occurs in less than 5 min, the sample temperature is lowered in 3°C steps until ignition does not occur. Thus, the ASTM minimum AIT is based on an ignition delay of approximately 300 s. With such a minimum AIT of 300°C, ignition might occur at 295°C in 420 s, for example (E = 167.4). This would have to be established by tests; Eq. (3-1) does not give an absolute minimum temperature at which ignition could take place.

In AIT tests, the gas mixture and surfaces of the heated vessels are substantially at the same temperature. A different situation occurs when a gas mixture is exposed to a hot surface, e.g., a heated wire or tube. Then only a portion of the mixture is in contact with the hot surface. The ignition-temperature values increase with a decrease in heat-source surface area, and the temperatures required for ignition by heated wires, rods, and tubes tend to be higher than those for autoignition.[5,6]

Concentration of Vapors

In a homologous series of organic compounds, the AIT decreases with increasing molecular weight, as shown in Table 3-1 for paraffin hydrocarbons.[7] Also, a branched compound will have a higher AIT than the corresponding straight-chain compound.

TABLE 3-1 Autoignition Temperatures of Paraffin Hydrocarbons in Air at 101.325 kPa[7]

	AIT, °C		AIT, °C
Methane	537	*n*-Heptane	223
Ethane	515	*n*-Octane	220
Propane	466	*n*-Nonane	206
n-Butane	405	*n*-Decane	208
Isobutane	462	*n*-Dodecane	204
n-Pentane	258	*n*-Hexadecane	205
n-Hexane	223		

Burgoyne[8] has attributed the sharp drop in AIT between butane and higher-molecular-weight paraffins to the appearance of cool flames with the higher-molecular-weight compounds.

For any given compound, the combustible-to-air ratio affects the AIT to some degree. Mixtures that are excessively fuel-rich or excessively fuel-lean generally ignite at higher temperatures than those at intermediate compositions.[6,9] In ASTM D2155, 0.07 mL of a liquid is injected into a 200-mL Erlenmeyer flask for determination of an AIT. Then 0.10 mL is tried. If this second test gives a lower AIT, the method is repeated using 0.12, then 0.15 mL, etc., until a minimum AIT is found. If the 0.10-mL sample has a higher AIT than the 0.07-mL sample, the procedure is performed with 0.05- then 0.03-mL samples. Thus, the ASTM D2155 method covers a wide range of fuel concentrations for determination of the minimum AIT. Normally, the minimum AIT occurs with 0.04 to 0.05 mL of the combustible liquid in a 200-mL flask, which is between the stoichiometric concentration and upper flammability limit.[9] With hexane as an example these quantities equal about 6 to 7% v/v at the AIT; the U and C_{st} for hexane are 7.5 and 2.16% v/v, respectively.

Environmental Effects

Setchkin[9] observed that over a vessel size range of 8 mL to 12 L the AIT becomes lower with increasing vessel size in flasks of similar construction. He attributed this effect to the smaller rate of heat loss per unit volume from the reacting medium with the larger flasks because of the smaller ratio of surface to volume. His data for toluene and methyl alcohol for flask sizes 35 mL to 12 L are shown in Fig. 3-1.

An increase in pressure usually decreases AITs, and a decrease in pressure raises AITs. Minimum AITs of mineral oils and kerosene in air at high and low pressures, respectively, are shown in Table 3-2.

Occasionally, processes are operated in oxygen-enriched or oxygen-depleted air. Usually, oxygen enrichment of air tends to decrease the minimum AIT. A decrease in O_2 percentage in air, on the other hand, increases the minimum AIT. The effect of O_2 percentage on the mini-

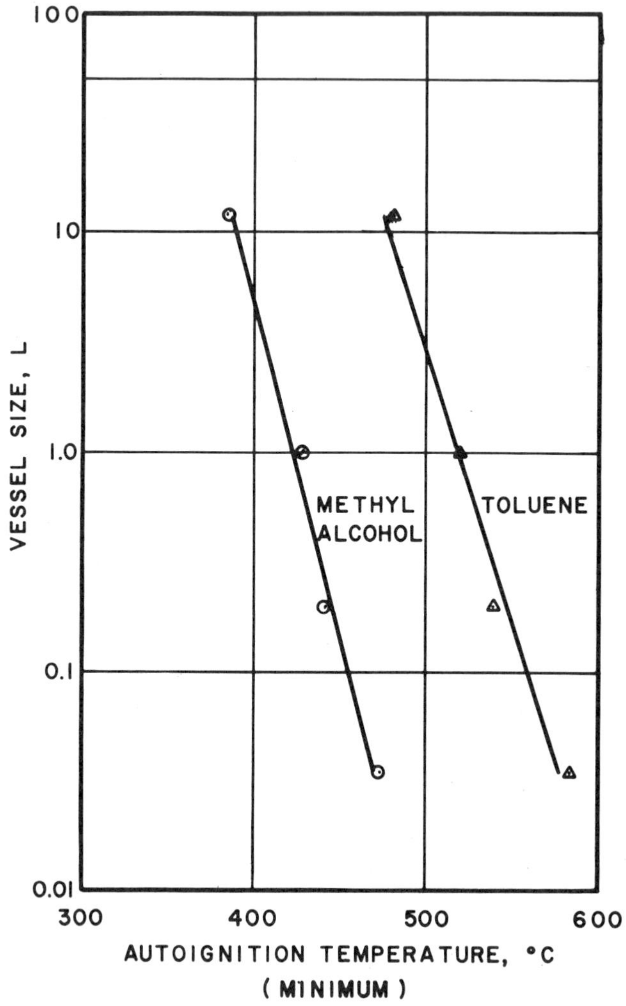

Fig. 3-1 Effect of vessel size on minimum autoignition temperature. (*Data of Setchkin.* [9])

mum AIT of JP-6 fuel at 101.325 kPa is shown in Fig. 3-2. (JP-6 is a kerosene-base fuel.)

Thus, the environmental effects just described highlight the profound influence they have on ignition. Also, low-temperature oxidation can result in cool flames, which may then grow into ignition (see Secs. 2-4 and 2-5).

Catalytic Material

In some explosions ignition may occur in equipment where the temperature is less than the minimum AIT, even considering the environmental effects cited in the previous section. Catalytic materials, such as metal oxides, can promote oxidation on their surfaces, leading to a high local temperature and subsequent ignition of the entire gas mixture. Hilado and Clark[13] investigated the effect of ferric oxide powder on the autoignition behavior of 21 combustible organic compounds. All the compounds that showed AIT lowering by ferric oxide powder had AIT values above 290°C in glass. Thus 290°C may well be the practical upper limit of AIT for all combustible organic compounds coming in contact with rusty iron or steel. Moreover, catalysts are used in many gas- and liquid-phase reactions. Some catalytic material may spread unknown throughout the process equipment. It can thus act as an omnipresent ignition source, lurking to ignite flammable mixtures, possibly developed if the composition of the feed gets out of control.

Flow Conditions

In an unconfined space or under flow conditions, heat must be transferred to the moving gas mixture to produce ignition. For laminar flow, Thiyagarajan and Hermance[14] found that when the heat-transfer time is more than the oxidation time to develop ignition, the ignition temperatures of various hydrocarbons are similar, i.e., about 800°C. Tests in the open by Husa and Runes[15] showed that the surface temperature required for ignition of hydrocarbons is about 750°C. Nevertheless, they cautioned that a hot surface below 750°C in these circumstances can become an indirect source of vapor ignition by first igniting paper, etc., blown against the surface. Moreover, ignition may occur at the minimum AIT if stagnant and flammable conditions develop unexpectedly, due to a power failure for instance.

3-2 Autooxidation

Autooxidation is the phenomenon of slow oxidation with accompanying evolution of heat leading to autoignition when the heat cannot be dissipated adequately. Only liquids with low volatility pose risks from this

TABLE 3-2 Minimum Autoignition Temperatures in Air at High and Low Pressures[10,11]

	Minimum AIT, °C				
	25 kPa	50 kPa	100 kPa	1 MPa	10 MPa
Mineral oils	. . .	. . .	350	250	200
Kerosene	593	464	229		

type of autooxidation; volatile liquids, such as toluene, evaporate quickly.

Thermal insulation wetted with oil has been the source of numerous plant fires in which the insulated hot surface was below the autoignition temperature.[16–19] The autooxidation temperature of Dowtherm A is reported to be as low as 260 to 288°C[17]; its published AIT is 621°C.[Dowtherm A is a heat-transfer medium and a eutectic mixture of diphenyl and phenyl ether (diphenyl oxide).] Insulation and matted fibers provide a large surface area for oxidation, preventing dissipation of the evolved heat and leading to higher temperatures until autoignition occurs. As with the classic oily-rag and wet-hay phenomena, some air is necessary for oxidation; too much air will remove heat. Ignition may occur soon

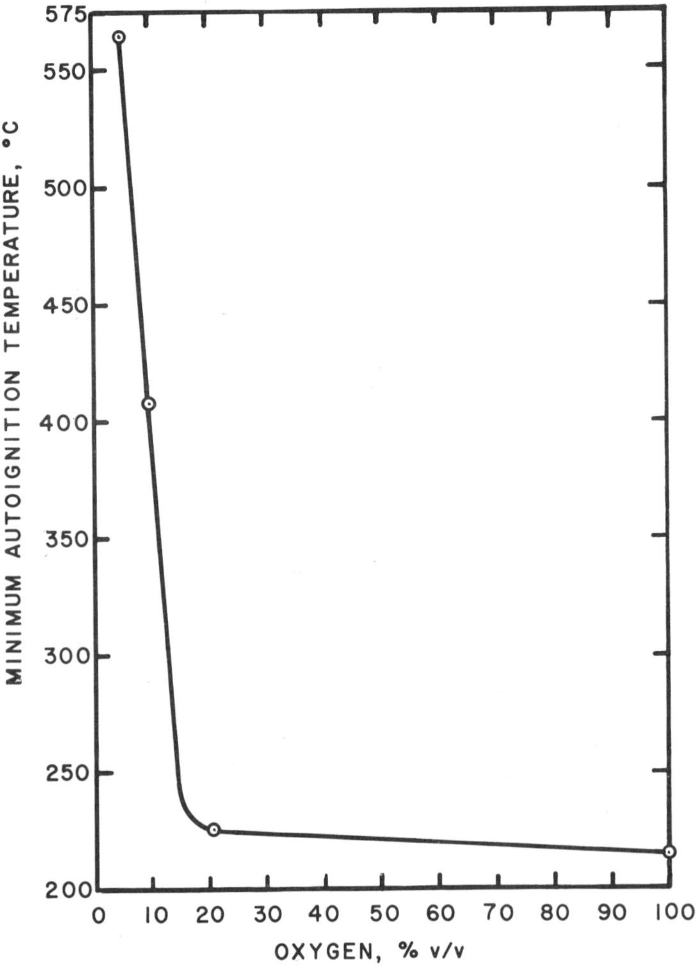

Fig. 3-2 Effect of oxygen on minimum autoignition temperature of JP-6 fuel–oxygen–nitrogen mixtures in a 2-L flask at 101 kPa abs. (Flames at 5% 0_2 did not propagate.) (*After Kuchta et al.* [12])

after the original soaking, or it may take months, depending upon the complex chemical and physical processes involved.

Several methods can be used to reduce wetting of insulation by oil:

1. Installation of all valves with stems in a horizontal position so that leaks will drip away from the insulation[18]

2. Installation of sheet-metal troughs to carry leakage from the insulation below[18]

3. Use of metal jacketing as covering, e.g., fire-resistant steel jacketings like Steel Jac, Tuf Jac and Steel-Cote.*

Insulation that is known to be wetted with oil should be removed promptly.[18,19]

3-3 Electrical Ignition

Article 500 of the National Electrical Code (NEC) divides fire and explosion hazards into three classes.[20] (API Recommended Practice for Classification of Areas for Electrical Installations in Petroleum Refineries, *API RP* 500A, is a guide of the American Petroleum Institute for petroleum refineries. *NFPA* 497 is the Recommended Practice for Classification of Class I Hazardous Locations for Electrical Installations in Chemical Plants.) *Class I* locations are areas in which flammable gases or vapors are or may be present in the air in quantities sufficient to produce explosive or ignitable mixtures. *Class II* locations are areas that are hazardous because of the presence of combustible dust. (*Class III* covers flammable fibers and flyings and will not be discussed in this text.) Each class is divided into two divisions, as follows:

Class I, Division 1

a. Location in which hazardous concentrations of flammable gases or vapors exist continuously, intermittently, or periodically under normal operating conditions or

b. In which concentrations of such gases or vapors may exist frequently because of repair or maintenance operations or because of leakage or

c. In which breakdown or faulty

Class I. Division 2

a. Location in which volatile flammable liquids or flammable gases are handled, processed, or used, but in which the hazardous liquids, vapors, or gases will normally be confined within closed containers or closed systems, from which they can escape only in case of accidental rupture or breakdown of such containers or systems, or in case of abnormal operation of equipment or

*Steel Jac, Tuf Jac, and Steel-Cote are registered trademarks of Childers Products Co., Insul-Coustic Birma Corp., and Preformed Metal Products Co., Inc., respectively.

operation of equipment or processes might release hazardous concentrations of flammable gases or vapors, and might also cause simultaneous failure of electric equipment

b. In which hazardous concentrations of gases or vapors are normally prevented by positive mechanical ventilation, and which might become hazardous through failure or abnormal operation of the ventilating equipment or

c. which is adjacent to a class I, division 1 location and to which hazardous concentrations of gases or vapors might occasionally be communicated unless such communication is prevented by adequate positive-pressure ventilation from a source of clean air and effective safeguards against ventilation failure are provided

Class II, Division 1

a. Location in which combustible dust is or may be in suspension in the air continuously, intermittently, or periodically under normal operating conditions in quantities sufficient to produce explosive or ignitable mixtures or

b. where mechanical failure or abnormal operation of machinery or equipment might cause such explosive or ignitable mixtures to be produced and might also provide a source of ignition through simultaneous failure of electric equipment, operation of protection devices, or from other causes or

c. In which combustible dusts of an electrically conducting nature may be present

Class II, Division 2

a. Location in which combustible dust will not normally be in suspension in the air or will not likely be thrown into suspension by the normal operation of equipment or apparatus in quantities sufficient to produce explosive or ignitable mixtures, but

(1) Where deposits or accumulations of such combustible dust may be sufficient to interfere with the safe dissipation of heat from electric equipment or apparatus or

(2) Where such deposits or accumulations of combustible dust on, in, or in the vicinity of, electric equipment might be ignited by arcs, sparks, or burning material from such equipment

Class I locations are subdivided into four groups specified in the NEC; class II locations are subdivided into three groups. The NEC groupings do not cover all chemicals, however.

Division 2 locations are normally not hazardous. Thus, they apply to plant areas that can become hazardous only in the event of accidental discharge of flammable materials from confined systems. Consequently, the electrical requirements are less stringent than for division 1 locations. Explosion-proof equipment is generally required for class I division 1 locations. (Motors and generators may be totally enclosed, with additional specifications given in Article 501-8 of the NEC.) Explosion-proof electric equipment is not gastight. A cast-metal electrical enclosure must be capable of withstanding a hydrostatic pressure of 4 times the maximum pressure of an internal explosion without rupture or permanent deformation. It must also prevent escape of flames and operate below the ignition temperature of the flammable material in the ambient environment. Essentially, the chief purpose of an explosion-proof enclosure is to prevent initiation of a fire or explosion in the ambient atmosphere. Normally, nonsparking equipment or apparatus that has make-or-break contacts immersed in oil or hermetically sealed is used in division 2 areas. This division 2 equipment can cause ignition only if it malfunctions at the same time a flammable concentration develops, an unlikely event considering the tiny probability of simultaneous occurrence of electrical failure with release of flammable materials. When nonsparking equipment is not available, explosion-proof equipment or electrical apparatus contained in explosion-proof housings is used in division 2 locations.

If the surface temperatures of lamps in class I division 2 locations reach surface temperatures exceeding 80 percent of the ignition temperature (in degrees Celsius) of the gas or vapor involved, fixtures must comply with class I division 1 requirements or "shall be of a type which has been tested and found incapable of igniting the gas or vapor if the ignition temperature is not exceeded."[20]

Apparatus permissible for use in class I locations is not necessarily permissible for class II. Devices used in division 1 locations with combustible dusts have to be dust-ignition-proof but not explosion-proof. This equipment is constructed so that dust entry is excluded and so that exterior dust cannot be ignited by sparks or heat generated inside the equipment. Alternatively, class II division 2 locations may be constructed with enclosures to minimize deposits or entrance of dusts and to prevent the emission of *incendive* sparks, i.e., sparks capable of igniting a substance.

Articles 500 to 503 of the NEC prescribe rules for the installation of electrical wiring and equipment in hazardous areas. Special care is required in installation, use, and maintenance. Maintenance principles have been enumerated by Short.[21]

There are alternatives for downgrading electrical requirements without jeopardizing safety:

1. Locating electrical equipment in less hazardous or nonhazardous areas[20]

2. Providing adequate positive-pressure ventilation from a source of

clean air in conjunction with effective safeguards against ventilation failure[20,22]

3. Employing intrinsically safe equipment.

Outdoor, well ventilated process areas are class I division 2.[23] Establishing the boundaries for division 2 areas is often difficult. Quantity of emission is not an important criterion, unless the emission is instantaneous. The rate of release together with wind and turbulence in the environment and specific gravity govern downwind concentrations; 1 Mg of a substance released over a period of 10 min will give higher concentrations at any point downwind than 2 Mg of the same material released during 50 min. Large release rates can give flammable concentrations at large distances from the source of the release, as discussed at greater length in Chap. 6. Nevertheless, extension of the boundary of a class I division 2 area to these large distances would not necessarily be warranted, particularly when nonelectrical ignition sources, such as hot surfaces, may exist there.

Intrinsically safe electrical equipment cannot release enough electrical energy to ignite a specific hazardous atmosphere. Article 500-1 of NEC states that

> Intrinsically safe equipment and wiring shall not be capable of releasing sufficient electrical or thermal energy under normal or abnormal conditions to cause ignition of a specific hazardous atmospheric mixture in its most easily ignited concentration. Abnormal conditions shall include accidental damage to any field-installed wiring, failure of electrical components, application of over-voltage, adjustment and maintenance operations, and other similar conditions.

The low energy requirements limit the use of intrinsically safe equipment to low-power devices, such as process-control instrumentation and communication equipment. It cannot be used for large motors or general lighting.[24] NFPA Standard 493 provides requirements for the construction and test of electrical apparatus or parts of such apparatus in which the circuits themselves are incapable of causing ignition in class I division 1 hazardous (classified) locations, in accordance with Articles 500 and 501 of the NEC.[25]

After areas have been classified in accordance with the NEC, it is particularly important to ensure that no other sources can be present to cause ignition. Otherwise, classification and the special equipment it requires could go for naught. The potential danger of static electricity from accidental releases of flammable materials should be considered, for example. Contaminated gases released into an area can be electrically charged. The pipe from which they issue and the objects upon which flammable gases and liquids can impinge, e.g., pipe hangers and metal jacketing over thermal insulation, should be grounded.

Minimum Electric-Spark Ignition Energy

Figure 3-3 shows how minimum ignition energy (MIE) depends on the concentration of a combustible in air. The lowest MIE occurs near the stoichiometric concentration, but the greater the molecular weight of the compound the more the MIE shifts to the higher stoichiometric fraction. The MIE minima are about 0.25 mJ at sea-level pressure. (Hydrogen = 0.017 mJ.[26]) These MIEs for hydrocarbons are based on the optimum spark-gap length of about 2.5 mm. Greater electrode spacing will result in higher energy requirements for ignition; the MIE for break sparks, occurring when a switch is opened, for example, is in the range of millijoules rather than fractions of a millijoule.

Figure 3-4 shows how oxygen and pressure affect MIEs in mixtures of propane, oxygen, and nitrogen. Increased O_2 drastically reduces MIEs; at 101.325 kPa the smallest MIE decreases from 0.26 mJ in air to 0.002 mJ in oxygen. A decrease in pressure increases MIEs. Halving the total pressure will increase the minimum energy by a factor of about 5.[27]

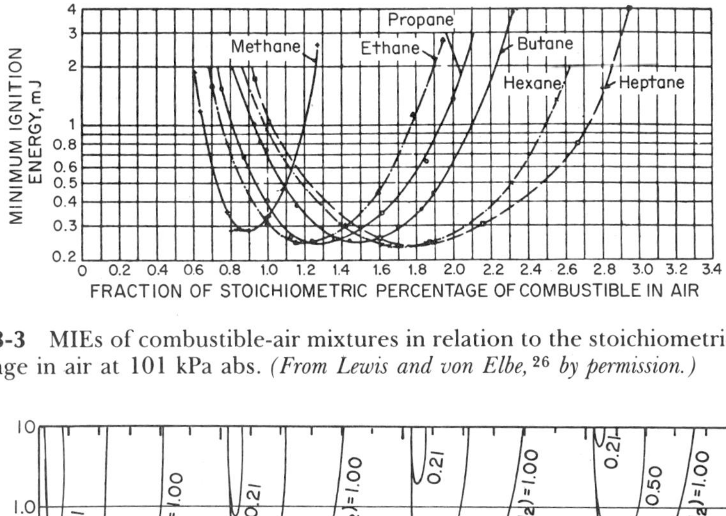

Fig. 3-3 MIEs of combustible-air mixtures in relation to the stoichiometric percentage in air at 101 kPa abs. *(From Lewis and von Elbe,[26] by permission.)*

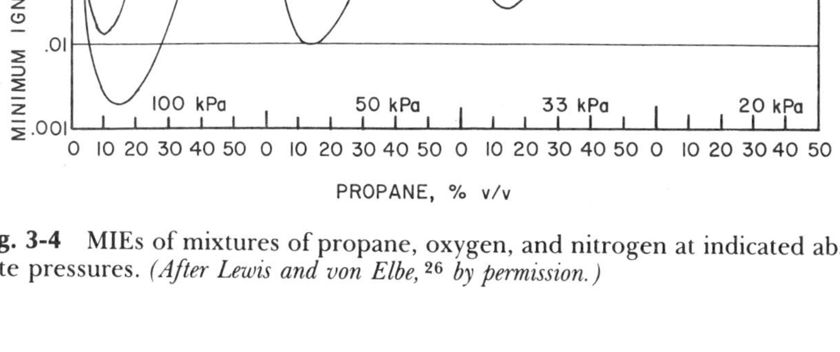

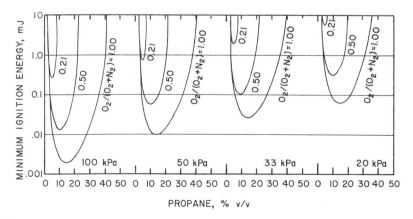

Fig. 3-4 MIEs of mixtures of propane, oxygen, and nitrogen at indicated absolute pressures. *(After Lewis and von Elbe,[26] by permission.)*

A person with a capacitance of 200 pF and charged to 15 kV could initiate a discharge of 22.5 mJ. The energy in an ordinary spark plug is 20 to 30 mJ. Thus, except in intrinsically safe electrical equipment, commonplace sparks and arcs can ignite flammable vapor and gas mixtures with some energy to spare. More energy is required to ignite dusts, however, as shown in Fig. 3-5.[28]

Static Electricity

Static electricity is caused by the contact and separation of materials, generally an electrical conductor and a nonconductor or two nonconductors. Separation results in the accumulation of a negative charge on one of the materials and an equal positive charge on the other. When the objects are separated, work must be done against the attractive force between the negative and positive charges. The potential (voltage) produced equals the amount of work which has to be done against this force.

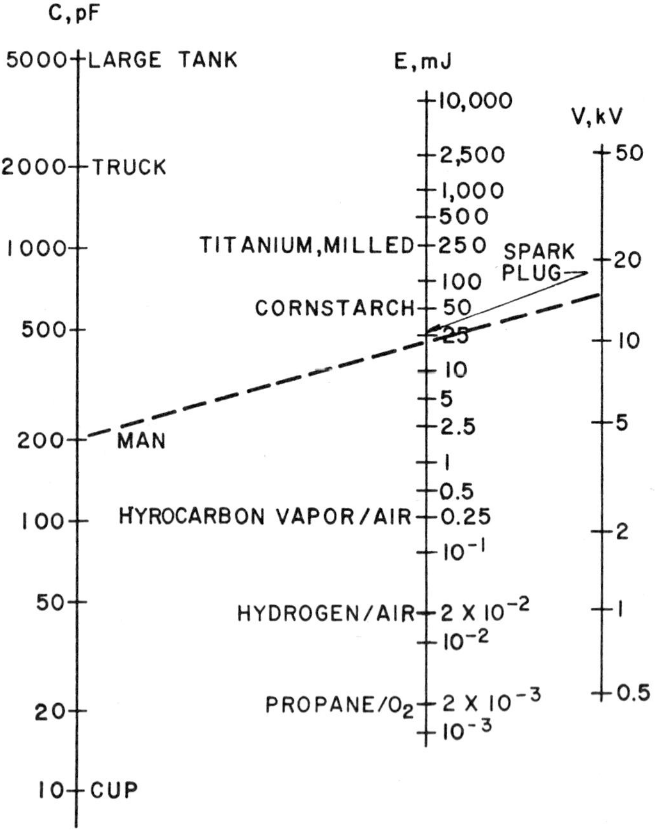

Fig. 3-5 MIEs for various gases and dusts in turbulence. Particle size <200 mesh, not necessarily of the same size distribution. *(Data from U.S. Bureau of Mines; after Owens,* [28] *by permission.)*

Charge is measured in coulombs, and the *capacitance* of an object is the ratio of the charge on the object to its potential. The capacitance depends on size and becomes greater as the object gets larger. With a given charge, the voltage gets higher as the capacitance gets smaller. The energy W stored in a charged capacitor is

$$W = 0.5CV^2 \times 10^{-3} = 0.5QV = \frac{500Q^2}{C} \tag{3-4}$$

where W = energy, mJ
C = capacitance, pF
V = potential difference, kV
Q = charge, μC

This energy is released as a spark from an insulated conductive body when the voltage becomes great enough. (The minimum sparking potential for charged electrodes is about 350 V, occurring at a spacing of only 0.01 mm.) Sparks from an equally charged nonconductor are less energetic and may contain only a portion of the stored energy. Such comparatively weak sparks are not likely to ignite dust clouds, but they can ignite flammable vapors or gases (see the discussion of particle size and static electricity in Sec. 8-2).

There are several general methods for controlling static electricity.

Bonding and grounding. Bonding means the equalization of potential between two conductive bodies by connecting them together with conductive wiring. Objects thus connected could still retain a charge relative to ground, for example. With grounding, a conductive object is connected to earth by a conductive wire; no. 8 or 10 AWG wire is about the minimum acceptable size. *Bonding and grounding of conductive equipment are the principal bulwarks against static-electricity hazards.* Wire in flexible hose should be grounded. Proper maintenance of the bond or ground wire is essential; a broken wire could become a charged conductor and itself become an ignition threat. Temporary connections can be made with clamps, but checks should be made to ensure that paint, etc., does not prevent proper electrical contact. A ground resistance of less than 1 MΩ is adequate for static grounding.[29]

Ionization. Ionization of air in contact with a charged body provides a conducting route through which the charge may dissipate. On a conducting sphere isolated in space, the distribution of charges is uniform. If the sphere is close to other objects, its charge distribution is distorted. If the object is not a sphere, the charge will concentrate on the surface with the smallest radius of curvature. A needlepoint has an almost zero radius of curvature. The electric field produced by the charge concentration on the point can become large enough to ionize air. A *static comb* is a metal bar equipped with rows of needlepoints or a wire wrapped with metallic tinsel. When a grounded static comb is close to an insulated charged body, ionization of the air surrounding the points of the comb will provide a conductive path over which the charge on the insulated body can leak away rapidly. Additional techniques of ionization are covered in Ref. 29.

Humidification. The conductivity of electrical nonconductors, such as plastics, paper, and concrete, depends on their moisture content. Relatively high moisture in these materials increases conductivity and thereby increases dissipation of static electricity. The moisture in these articles is in equilibrium with the moisture in the atmosphere, as measured by the relative humidity of the air. (Relative humidity is the ratio of actual water vapor in the air to the maximum amount of water vapor air can hold at a given temperature.) Low relative humidity permits moisture in the materials to evaporate, decreasing conductivity and the drainage of static electricity. With high relative humidities of 60 to 70 percent a microscopic film of moisture covers surfaces, making them more conductive. Sometimes humidification is practiced to decrease the hazard of static electricity, but it is not a general method of control.

Generation and Control of Static Electricity

Static electricity can be generated in industry in several ways. To accumulate charges, however, the dissipation of charges must be slower than generation. Several common ways by which static electricity is produced are as follows.

Moving nonconductive power and conveyor belts. V belts are not as susceptible to hazardous static generation as flat belts, and it is generally considered that their risk from the standpoint of static electricity is small.[29] Otherwise, the best method of control is to use conductive belting. (Slipping belts cause heat buildup by friction and are a cause of fires.)

Flow of fluids through pipes into tanks or other containers. When liquids flow through closed metal pipes, static electricity is not a hazard. It may become a hazard, however, when liquids are pumped into tanks. Charges produced in the liquid during pumping can accumulate on the surface of the liquid and cause sparking between the liquid surface and tank or projection into the tank. Although piping and the tank should be grounded, the grounding does not necessarily eliminate this danger for poorly conductive flammable liquids. The degree of hazard depends on how fast the liquid loses its charge, as expressed by relaxation time. *Relaxation time* is the time it takes for 63 percent of the charge to leak away from a charged liquid through a grounded conductive container. It is 144 percent of the half-value time $\tau_{1/2}$.[28]

$$\tau_{1/2} = \frac{6.15\ \epsilon}{k} \tag{3-5}$$

where $\tau_{1/2}$ = half-value time, s (time required for magnitude of free charge to decay to one-half its initial value)
ϵ = relative dielectric constant of liquid, dimensionless
k = conductivity of liquid, pS/m (1 pS/m = 10^{-8} μS/cm; siemens were formerly called mhos)

A volume resistivity of 10^{14} Ω·cm is equivalent to a conductivity of 1 pS/m.

A half-value time of less than 0.012 s is reported not to constitute a hazard.[30] Toluene is a notoriously poor conductor, with a conductivity of 1 pS/m and a half-value time of 14.6 s. It has been involved in more than 75 percent of the paint fires to come to the attention of Freriks.[31] Also, the saturated vapor pressure of toluene at normal room temperature (20°C) is 2.89% v/v. This is slightly richer than the stoichiometric concentration in air of 2.28% v/v. Accordingly, as noted in the previous section on MIE, it is one of the easiest concentrations to ignite. Moreover, as discussed in greater detail in Chap. 4, the explosion effects are greatest at this concentration.

Relaxation and half-value times for some common flammable liquids are given in Table 3-3.

Ultra-low conductivities of less than about 0.1 pS/m may not constitute a static hazard because of the lack of ions to produce dangerous charges. Nevertheless, contamination could result in raising the conductivity enough to give a still low and potentially hazardous value.

Filters in pipelines greatly increase the generation of static electricity. In one test it was reported that the charge development in aircraft fueling tests was 10 to 200 times more with a filter than

TABLE 3-3 Relaxation and Half-Value Times near Normal Room Temperature

	Conductivity, pS/m	Dielectric constant	Half-value time, s	Relaxation time, s
Acetaldehyde	0.17×10^9	21.0	0.76×10^{-6}	1.09×10^{-6}
Acetone	5.9×10^6	20.7	21.6×10^{-6}	31.1×10^{-6}
n-Hexane	0.1×10^{-3}	1.89	0.116×10^6	0.167×10^6
Methyl alcohol	43×10^6	32.6	4.61×10^{-6}	6.64×10^{-6}
Methyl ethyl ketone	10×10^6	18.5	11.4×10^{-6}	16.4×10^{-6}
Toluene	1.0	2.38	14.6	21.0

without one.[32] Also, settling out of a conductive phase through a nonconductive phase, such as water in oil, greatly increases the hazards of generation of static electricity. Thus, handling an emulsion could be more hazardous than handling a single-phase system. Agitation or any other operation that breaks up the flammable liquid into droplets increases static generation. OSHA regulations (Code of Federal Regulations 1910.106) require fill pipes to terminate within 152 mm (6 in) of the bottom of a tank for many class I-B and class I-C liquids. (Class I-B includes liquids having flash points below 22.8°C and a boiling point at or above 37.8°C. Class I-C liquids include liquids having flash points at or above 22.8°C and below 37.8°C.)

The conductivity of flammable liquids can be artificially raised to generally safe levels of 50 to 100 pS/m by additives. These conductivity improvers reduce electrostatic hazards by increasing the rate of dissipation of electrostatic charge without necessarily reducing the amount of electrostatic charge generated. For example, 0.3 to 1 mg/L of Stadis* 450 will give conductivities of 100 pS/m or more in most hydrocarbon solvents. Low pumping rates also reduce buildup of static electricity. With nonconductive linings or vessels, some type of grounded insert or probe should be installed in the vessel.[33] Care must be taken to prevent breakage of the grounding connection; otherwise the probe will become an insulated conductor with high sparking potential. In addition, all sampling probes and containers preferably should be nonconductive; a lost conductive object floating in a tank could cause sparking when it approaches the tank wall. Moreover, in some cases it may be prudent to deplenish oxygen with inert gas, as discussed in Chap. 2.

Accumulation of charges on personnel. A person can accumulate dangerous charges up to about 20 kV maximum when relative humidity is low. As shown in Fig. 3-5, the associated energy can ignite most flammable vapors and gases, and it may ignite ignition-sensitive dusts. An ungrounded person may accumulate significant charges when handling powders. Such charges will not ignite most powders, but they can ignite flammable vapors that sometimes are associated with powder handling, e.g., when a powder is manually dumped into a vessel containing a flammable liquid.

Grounding is the usual method of controlling charge accumulation on personnel. For static control personnel should wear leather shoes, which are conductive, and not insulating rubber footwear.

*Stadis is the registered trademark of E. I. du Pont de Nemours & Company for its antistatic additives.

Legstats or Wriststats (Walter G. Legge, Co., Inc.) are available for persons who are not wearing conductive footwear. (Legstats provide a conductive path from the leg to the floor. Wriststats are conductive straps worn on the wrist with a connecting 0.91-m-long cable that clamps to grounded equipment.) It is helpful for persons to touch a grounded object to dissipate charges before entering potentially hazardous locations.

To provide proper grounding of personnel, flooring must be conductive too. The resistance of a conductive floor should be less than 1 MΩ as measured between two electrodes placed 0.91 m apart at any point on the floor.[34] (As an additional precaution against electric shock, hospital operating rooms have resistances of more than 25 kΩ measured in the same manner.[34]) It may sometimes be necessary to install a conductive plate to provide proper grounding of personnel in particular locations. Care must be exercised to prevent accumulation of foreign material, e.g., powders and floor wax, from turning an ostensibly conductive floor or plate into a nonconductive one. Conductivities should be checked periodically.

Flow of gases from nozzles and stacks. Pure gases do not generate significant static electricity in transmission through pipes and ducts. Gases contaminated with rust particles or with liquid droplets, however, do produce static, but this is not a problem in a closed, grounded piping system. If these gases impinge on an ungrounded conductive object, dangerous charges can accumulate on that object. "Wet" steam, i.e., steam with droplets of water, can develop charges. If these touch an ungrounded conductor, electrification of that ungrounded object can occur.

Flammable gases often ignite when they are discharged to air from stacks during thunderstorms. (Even without a direct lightning hit, the electric field, i.e., potential gradient, is so strong that charges from the ground are attracted to the tip of a grounded stack and flow out as a corona discharge.) Steam is often injected into a stack in these circumstances to inhibit ignition. Steam may not extinguish an ignited release, however, unless it is supplied in prodigous quantities. (Actually, steam is supplied to flares at the rate of about 0.3 kg per kilogram of flammable gas to promote smokeless combustion.) Dry hydrogen, acetylene, and occasionally other gases often ignite when they are discharged to air in normal weather. Apparently, the electric field developed by ejection of these charged gases can develop enough of a potential gradient to cause ignition by corona discharge; as noted previously, the MIE of hydrogen is only about 0.02 mJ. NFPA Code 78 covers lightning protection.[35] A NASA toroidal ring (Fig. 3-6) is reported to prevent unwanted discharge and subsequent ignition at a vent-stack outlet.[36] Outside diameter *d* of the tubing forming the toroidal ring depends on the diameter of the vent stack, as follows:

D, mm	d, mm
<203	12.7
203–305	19.0
>305	25.4

3-4 Friction

Sparks can occur when two hard materials come into forced contact with one another. Only glancing blows produce friction sparks. (MIEs of electric sparks are not directly relatable to ignitions by frictional impact.) If an accidental release occurs, ignitions may happen outside or inside of process equipment. Hand and mechanical tools are the most likely sources of friction sparks outside equipment.

The need for nonsparking hand tools is often a controversial subject. Risinger indicated that he knew of only one case of vapor ignition by hand tools.[37] Even in that case "sparkproof" tools were being used. Threshold limit values (TLVs) for safe industrial exposure are far below lower flammability limits; it is extremely unlikely that a person will be using hand tools in a flammable atmosphere, and it is prudent to control the atmosphere rather than the tools. (Plants manufacturing explosives often require "spark proof" tools.) If it is necessary on rare occasions to eliminate potential sparking from hand tools categorically, the surface to be struck may be wetted with water or heavy oil.[37] Also, when tested, tools fabricated from manganese bronze, phosphorus bronze, aluminum bronze, commercial brass, aluminum, and beryllium copper did not ignite a flammable atmosphere of gasoline vapor.[38] Aluminum, copper, or

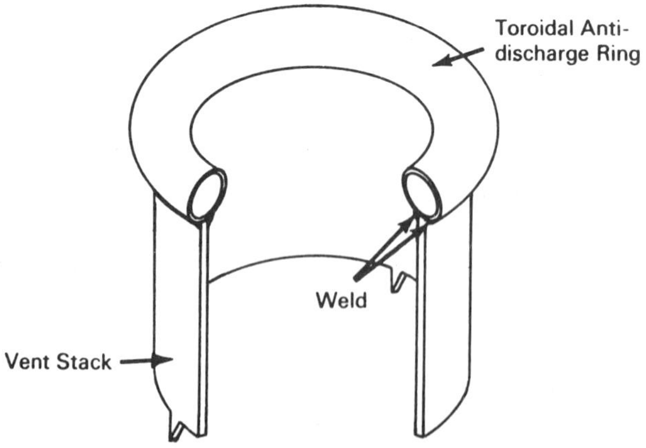

Fig. 3-6 Toroidal ring for prevention of gas ignition at vent-stack outlet. *(From Ref. 36.)*

bronze metal tools are required in manufacturing plants for aluminum or magnesium powder.[39] (Note comments on aluminum below.) Mechanical tools, such as pneumatic chisels, may cause incendive sparks, particularly if one of the surfaces becomes heated.[37,38] Operation of equipment in the dark in a test will reveal whether sparks are generated but not whether they are incendiary. (Hot spots first become visible to the eye between 520 and 570°C.[40])

Impact upon rusty steel coated with aluminum paint or frictional, i.e., glancing, impact of aluminum on rusty steel can produce incendiary sparks. In the former case the pipe has to have been preheated to produce incendive material.[37,41] A thermite reaction occurs:

$$2Al + Fe_2O_3 \rightarrow Al_2O_3 + 2Fe$$

$$\Delta H = -15.56 \text{ MJ/kg Al}$$

The risk of ignition by frictional impact with aluminum is considerably less with magnesium-free aluminum; the threshold incendive energies for magnesium-free and 1% Mg aluminum to produce incendiary sparks against rusty steel are 393 and 95 J, respectively, (Impact angles = 45°.[42]) Accident Case History 2161 of the Manufacturing Chemists' Association describes a case in which a thermite reaction, caused by friction of a steel valve turning on small threads of aluminum that had galled in a coupling, was a possible source of ignition for solvent vapors.

A thermite reaction also occurs between titanium and steel. The hazard of titanium to steel is reported to be great in a drop-weight test in a 6% methane-air mixture.[38] This reaction is

$$3Ti + 2Fe_2O_3 \rightarrow 3TiO_2 + 4Fe$$

$$\Delta H = -8.094 \text{ MJ/kg Ti}$$

In equipment interior, tramp metal entering a grinding mill or similar equipment has caused many explosions.[43] Entry of tramp metal can be impeded by installation of a magnetic separator at a suitable point in the feed to the mill.

Rubbing of materials can produce sufficient heat to ignite dusts and flammable gases. Ignitions of flammable gases in tests by rubbing between metals have occurred.[41] Local hot spots will occur more readily with materials having low thermal conductivity. Friction in bearings may generate sufficient heat to ignite dust; it is important that bearings be properly lubricated and maintained. Also, an overload switch on mill and screw feeder motors, for example, will help prevent hot surfaces if the mill or feeder becomes overloaded.

3-5 Compression

If a gas is compressed rapidly, its temperature will increase. Autoignition may occur if the temperature of the compressed gas becomes high enough. Explosions have occurred in air compressors due to autoignition of lubricating oils. Also, an advancing piston of high pressure gas can compress and heat trapped gas ahead of it. The temperature rise from adiabatic compression of a perfect gas is given by

$$\frac{T_2}{T_1} = \left(\frac{P_2}{P_1}\right)^{(\kappa - 1)/\kappa} \tag{3-6}$$

Compressed-gas temperatures for air and isobutane are shown in Table 3-4 for an initial temperature of 25°C. (Validity of the perfect-gas law is assumed for illustrative purposes.)

Substances with relatively low ratios of heat capacities are heated less. Thus, for example, if isobutane were being compressed from 101.325 kPa and air accidentally entered the inlet gas, ignition could develop with a compression ratio of only about 23.5 (see Table 3-1). Moreover, the adiabatic compression of small bubbles of gas trapped in a liquid by impact or other means may cause high enough temperature to initiate an explosion in a liquid explosive.[40]

TABLE 3-4 Gas Temperatures from Adiabatic Compression

Compression ratio, P_2/P_1	Air, °C (κ = 1.40)	Isobutane, °C (κ = 1.11)
5	199	77
10	302	101
20	429	120
23.5	462	134
50	639	166
100	838	197

References

1. National Fire Protecting Association, A Hazard Study, *NFPA* HS-9, Boston, 1974.

2. Semenov, N. N., *Some Problems in Chemical Kinetics and Reactivity,* vol. 2, Princeton University Press, Princeton, N.J., 1959.

3. Jost, W., *Explosion and Combustion Processes in Gases,* McGraw-Hill, New York, 1946.

4. Penner, S. S., and B. P. Mullins, *Explosions, Detonations, Flammability and Ignition,* Pergamon, New York, 1959.

5. Kuchta, J. M., A. Bartkowiak, and M. G. Zabetakis, "Hot Surface Ignition Temperatures of Hydrocarbon Fuel Vapor-Air Mixtures," *J. Chem. Eng. Data,* vol. 10, no. 3, pp. 282–288, July 1965.

6. Kuchta, J. M., R. J. Cato, and M. G. Zabetakis, "Comparison of Hot Surface and Hot Gas Ignition Temperatures," *Combust. Flame,* vol. 8, pp. 348–350, December 1964.

7. Zabetakis, M. G., "Flammability Characteristics of Combustible Gases and Vapors," *U.S. Bur. Mines Bull.* 627 (*USNTIS* AD701 576), 1965.

8. Burgoyne, J. H., "Principles of Explosion Prevention," *Chem. Process Eng. (Lond.),* vol. 42, no. 4, pp. 157–161, April 1961.

9. Setchkin, N. P., "Self Ignition Temperatures of Combustible Liquids," *Res. Pap.* 2516, J. Res. Natl. Bur. Stand., vol. 53, no. 1, pp. 49–66, July 1954.

10. Zabetakis, M. G., G. S. Scott, and R. E. Kennedy, "Autoignition of Lubricants at Elevated Pressures," *U.S. Bur. Mines Rep. Invest.* 6112, 1962.

11. Kuchta, J. M., S. Lambiris, and M. G. Zabetakis, "Flammability and Autoignition of Hydrocarbon Fuels under Static and Dynamic Conditions," *U.S. Bur. Mines Rep. Invest.* 5992, 1962.

12. Kuchta, J. M., A. Bartkowiak, and M. G. Zabetakis, "Autoignition of Hydrocarbon Jet Fuel," *U.S. Bur. Mines Rep. Invest.* 6654, 1965.

13. Hilado, C. J. and S. W. Clark, "Discrepancies and Correlations of Reported Autoignition Temperatures," *Fire Technol.,* vol. 8, no. 3, pp. 218–227, August 1972.

14. Thiyagarajan, R., and C. E. Hermance, "Prediction of Ignition Conditions for Flammable Mixtures Drifting over Heated Planar Surfaces," *Combust. Inst. Jt. Spring Meet. Southwest Res. Inst., San Antonio, April 21–22, 1975.*

15. Husa, H. W., and E. Runes, "Q. How Hazardous Are Hot Metal Surfaces?," *Oil Gas J.,* vol. 61, pp. 180 and 182, Nov. 11, 1963.

16. Petkus, J. J., "Oily Insulation Can Cause Plant Fires," *Hydrocarbon Process. Pet. Refiner,* vol. 42, no. 11, p. 251, November 1963.

17. Albrecht, A. R., and W. F. Seifert, "Accident Prevention in High Temperature Heat Transfer Fluid Systems," *Chem. Eng. Prog. 4th Loss Prev. Symp. Atlanta, 1970,* pp. 67–88,

18. Manufacturing Chemists' Association, *Case Histories of Accidents in the Chemical Industry,* vol. 1, *Accident Case Histories 1–596,* Washington, 1962.

19. Coffee, R. D.: Discussion in Ref. 17.

20. National Fire Protection Association, National Fire Codes, *NFPA* 70, Boston.

21. Short, W. A., "Electrical Safety in Process Plants . . . Electrical Equipment for Hazardous Locations," *Chem. Eng.,* vol. 79, no. 9 pp. 59–64, May 1, 1972.

22. National Fire Protection Association, Standard for Purged and Pressurized Enclosures for Electrical Equipment in Hazardous Locations, *NFPA* 496, Boston.

23. LeVine, R. Y., "Electrical Safety in Process Plants . . . Classes and Limits of Hazardous Areas," *Chem. Eng.,* vol. 79, no. 9, pp. 51–58, May 1, 1972.

24. Hickes, W. F., "Electrical Safety in Process Plants . . . Intrinsic Safety," *Chem. Eng.,* vol. 79, no. 9, pp. 64–66, May 1, 1972.

25. National Fire Protection Association, Standard for Intrinsically Safe Apparatus for Use in Class I Hazardous Locations and Its Associated Apparatus, *NFPA* 493, Boston.

26. Lewis, B., and G. von Elbe, *Combustion, Flames, and Explosions of Gases,* 2d ed. Academic, New York, 1961.

27. Roth, E. M., "Space-Cabin Atmospheres, pt II: Fire and Blast Hazards," *NASA Spec. Pub.* 48, 1964.

28. Owens, J. E., Static Electricity, *ISA Monogr.* 110, pp. 113–127, 1965.

29. National Fire Protection Association, Recommended Practice on Static Electricity, *NFPA* 77, Boston.

30. Klinkenberg, A., "Laboratory and Plant-Scale Experiments on the Generation and Prevention of Static Electricity," *Proc., Annu. Meet. Am. Pet. Inst., 1957.*

31. Freriks, R. D., "Safety by Design," *Off. Dig. Fed. Soc. Paint Technol.,* vol. 37, pp. 436–438, April 1965.

32. Harris, D. N., G. Karel, and A. L. Ludwig, "Electrostatic Discharges in Aircraft Systems," *Proc. Annu. Meet. Am. Pet. Inst., 1961.*

33. Dorsey, J. S., "Static Sparks: How to Exorcize 'Go Devils,' " *Chem. Eng.,* vol. 83, no. 19, pp. 203–205, Sept. 13, 1976.

34. National Fire Protection Association, Standard for the Use of Inhalation Anesthetics, National Fire Codes, *NFPA* 564, Boston.

35. National Fire Protection Association, Lightning Protection Code, *NFPA* 78, Boston.

36. *NASA Tech. Brief* 67-10098, April 1967.

37. Risinger, J. L., "Fire Protection Handbook: Part 5," *Hydrocarbon Process. Pet. Refiner,* vol. 41, no. 5, pp. 209–211, May 1962.

38. Bernstein, H., and G. C. Young, "Sparking Characteristics of Metals Used in Tools," *Mat. Des. Eng.,* vol. 52, no. 7, pp. 104–105, December 1960.

39. National Fire Protection Association, Standard for the Manufacture of Aluminum or Magnesium Powder, *NFPA* 651, Boston.

40. Bowden, F. P., and A. D. Yoffe, *Initiation and Growth of Explosions in Liquids and Solids,* Cambridge Monograph on Physics, Cambridge University Press, London, 1952.

41. Powell, F., "Ignition of Gases and Vapors," *Ind. Eng. Chem.,* vol. 61, no. 12, pp. 29–37, December 1969.

42. Downing, A. G., "Frictional Sparking of Cast Magnesium, Aluminum, and Zinc," *Natl. Fire Prot. Assoc. Q.,* vol. 57, no. 3, pp. 235–245, January 1964.

43. Anon., Dust Explosions in Factories, *Health Safety Work Booklet* 22, Department of Employment and Productivity, H. M. Stationery Office, London, 1970.

4

Explosion Pressure

In previous chapters, flammability limits and ignition sources were discussed. By operating outside of the range of flammability or by depleting oxygen, explosions may be prevented. Also, all reasonable measures should be made to exclude potential ignition sources. Nevertheless, situations may occur where it is not feasible to base safety on preventive measures. Consequently, protection facilities have to be provided, and they are discussed in Chap. 5. The pressure effects of an explosion must be known beforehand, however. Thus, the effects of an explosion (deflagration), i.e., explosion pressure and rate of explosion-pressure rise, are discussed in this chapter. Transition to detonation and blast effects is also covered.

4-1 Maximum Explosion (Deflagration) Pressure in Unvented Vessels

When an ordinary explosion (deflagration) occurs, the maximum pressure developed in a closed vessel depends upon the initial pressure, change in moles of gas, and change in temperature. Thus,[1]

$$P_m = \frac{P_i n_f T_f}{n_i T_i} \tag{4-1}$$

With complete combustion of propane (C_3H_8) in air, for example, there is a negligible change in moles of gas

$$C_3H_8 + \underbrace{5O_2 + 18.8N_2}_{\text{Air}} \rightarrow 3CO_2 + 4H_2O + 18.8N_2$$

$$n_i = 24.8$$
$$n_f = 25.8$$

Therefore, explosion pressure develops principally from an increase in temperature in the combustion process. Maximum measured absolute explosion pressure in air ordinarily is about 8 times the initial absolute pressure, occurring at a composition slightly richer than the stoichiometric concentration in air, as shown in Fig. 4-1 for propane (C_{st} = 4.02% v/v).[2,3] Pressures are not additive in mixtures; peak explosion pressure occurs near the stoichiometric concentration of the mixture in air.

Explosion pressures within *BEC* in Figs. 2-1 and 2-2 will also be greatest near the stoichiometric concentration. With depleted oxygen, maximum explosion pressures decrease along C_{st} toward E and can be estimated by the relationship

$$P_{mn} \approx \frac{P_{ma} C_{st,n}}{C_{st,a}} \tag{4-2}$$

where P_{ma} = maximum explosion pressure with air, kPa abs = 818.4 kPa abs for propane with T_i = 300 K

P_{mn} = maximum explosion pressure for specified oxygen with fuel-air-nitrogen mixture, kPa abs

$C_{st,n}$ = stoichiometric concentration for same specified oxygen, % v/v

$C_{st,a}$ = stoichiometric concentration in air, % v/v

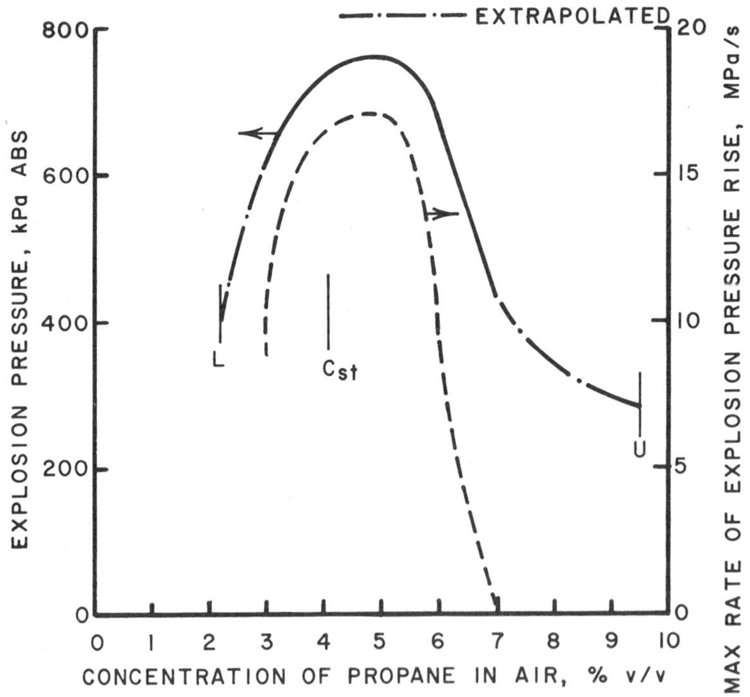

Fig. 4-1 Explosion pressure and rate of explosion-pressure rise for propane in air in a closed 10-L chamber at 65.5°C and an initial pressure of about 100 kPa abs. *(Based on Eastman Kodak Company data in NFPA 68, 1974.)*

For propane at E in Figs. 2-1 and 2-2

$$P_{mn} \approx \frac{(818.4)(2.20)}{4.02} \approx 448 \text{ kPa abs}$$

It is often necessary to determine the maximum pressure in an enclosure if only a pocket of material explodes. For a given amount of combustible material, the worst case will be stoichiometric conditions, and a hypothetical volume for those conditions should be calculated. Pressure may be equilibrated throughout the entire volume of the enclosure to obtain the maximum pressure in it, as shown by the following example.

Example

Inadvertently 0.1 m^3 of propane at 27°C (0.18 kg) is introduced into a 10-m^3 vessel and ignited. For propane, maximum explosion pressure occurs at 5% v/v in air. Thus, 717.1 kPa gage (818.4 kPa abs) theoretically could occur in a $0.1/0.05 = 2$ m^3 space. Maximum pressure in the 10-m^3 enclosure would be $(2/10)(717.1) \approx 143.4$ kPa gage. (Note that mixing the 0.1 m^3 propane in the entire 10 m^3 volume would give 1% v/v propane, i.e., below L, and no explosion pressure.)

The amount of gas to cause damaging pressure from ignition of a pocket of the gas can be estimated in a similar way. Assume that a 100-m^3 furnace, in which a mixture of 5% propane in air is burned, will be damaged at a pressure of 35 kPa gage. Explosion of only a

$$\frac{(35 \text{ kPa gage})(100 \text{ m}^3)}{717.1 \text{ kPa gage}} = 4.9\text{-m}^3$$

pocket of the 5% propane-air mixture will develop 35 kPa gage in the furnace. At a feed rate of 250 m^3/min of the mixture, delayed ignition of only $(4.9)(60)/250 = 1.18$ s can develop the damaging 35 kPa gage overpressure.

4-2 Rate of Explosion-Pressure Rise in Unvented Vessels

Explosion venting is a protection method discussed in Chap. 5. The rate of explosion-pressure rise $(dP/dt = r)$ is a key parameter in determining venting area (t = seconds). The rate of explosion-pressure rise for a mixture of 9.4% v/v methane in air is shown in Fig. 4-2 ($C_{st} = 9.5\%$ v/v). There is only a small pressure rise during the incipient stage of an explosion. Thus, in Fig. 4-2, pressure rises by only 101 kPa in the first 70 ms. Then, during an equivalent time to maximum explosion pressure, the pressure rises another 517 kPa. Experimental maximum dP/dt, r_m, occurs near the end of the explosion, i.e., when the flame has nearly reached the vessel wall.

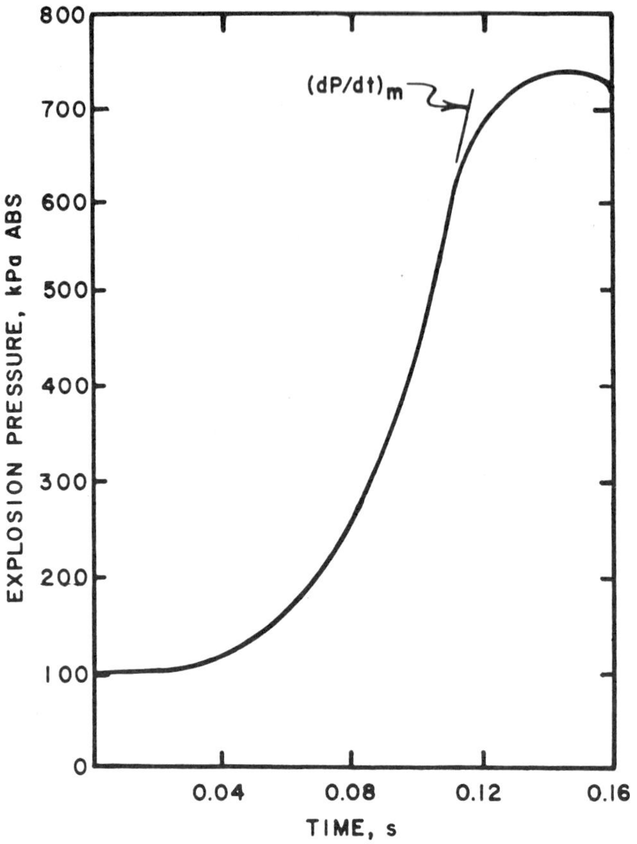

Fig. 4-2 Experimental explosion pressure for nonturbulent 9.4% v/v methane–90.6% v/v air mixture in a closed 28-L cubical vessel. (*After Nagy et al.*[4])

The concentration of the combustible affects the rate of explosion-pressure rise. The r_m occurs at a concentration slightly richer than the stoichiometric composition in air. A plot of r_m vs. concentration of propane is shown in Fig. 4-1. (Value of r_m in Figs. 4-1 and 4-3 may differ with other tests and is principally for description.)

4-3 Environmental Effects on Explosion Pressure in Unvented Vessels

The previous discussions of P_m and dP/dt have been based on a pressure of 101.325 kPa abs and normal atmospheric temperature. (Figure 4-1 is based on 65.5°C, however.) Also, the measured data were obtained in small vessels compared with the size of process vessels. Several environmental conditions affect P_m and dP/dt and are discussed in the following sections.

Temperature

The maximum explosion pressure decreases as temperature increases because of the smaller mass of material at the higher temperatures.[2,4] The maximum explosion pressure of propane (5% v/v in air) decreased from 818.4 kPa abs at 27°C to 597.8 kPa abs at 204°C.[2] On the other hand, r_m increases as temperature increases because burning velocity increases with an increase in initial temperature. Plots of both P_m and r_m vs. temperature for 5% propane in air are shown in Fig. 4-3.

Initial Pressure

Equation (4-1) indicates that P_m depends upon P_i. The relationship that peak explosion pressure is about 8 times the initial pressure is also

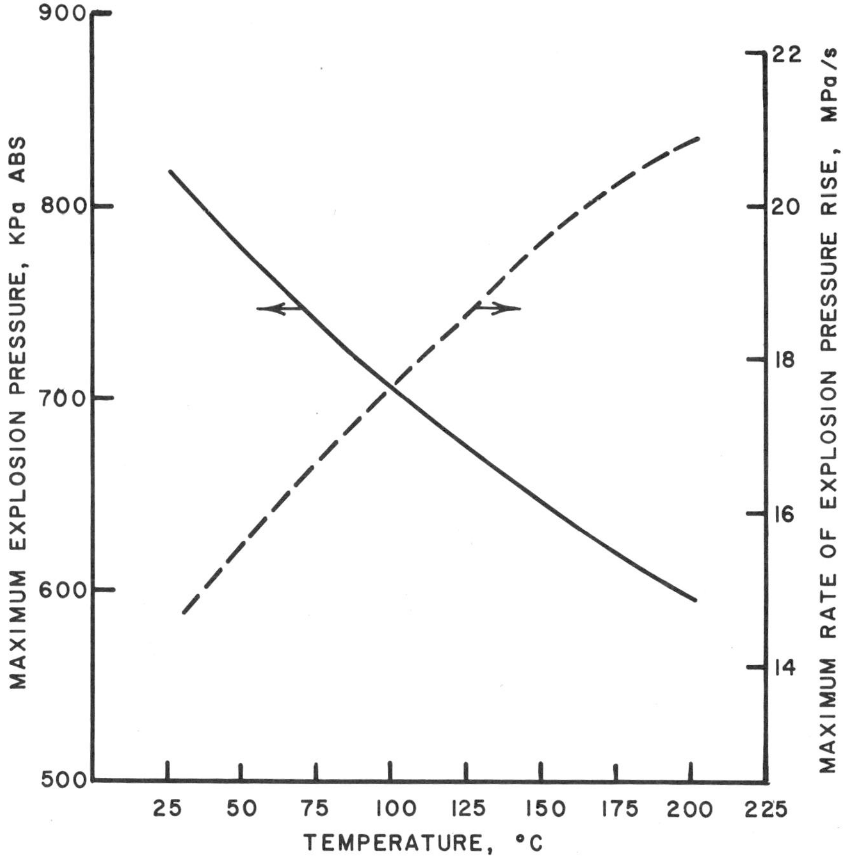

Fig. 4-3 Effect of temperature on maximum explosion pressure and maximum rate of explosion-pressure rise for 5% propane in air in a closed 10-L chamber at an initial pressure of about 100 kPa abs. *(Based on Eastman Kodak Company data in NFPA 68, 1974.)*

valid at elevated pressure.[4,5] This is illustrated in Table 4-1 for 5% v/v propane in air.

The maximum rate of explosion-pressure rise r_m increases linearly with initial pressure.[4,5] Thus, for example, for propane (C_3H_8) at 5% v/v and hydrogen (H_2) at 40% v/v in air in a 32-L vessel at atmospheric temperature[5]

$$r_{m,\,C_3H_8} = 0.63p_i + 17.2 \qquad p_i = 0 \text{ to } 310 \tag{4-3}$$

$$r_{m,\,H_2} = p_i + 68.9 \qquad p_i = 0 \text{ to } 310 \tag{4-4}$$

Vessel Geometry

Volume and shape. The maximum explosion pressure p_m is not significantly affected by the volume or shape of a vessel.[4,6] Where significant heat losses occur, however, as in equipment with a large length-to-diameter ratio, lower maximum pressure will result. Conversely, the maximum rate of explosion-pressure rise r_m is greatly affected by the volume V of a vessel. With a given compound and for similar vessel shapes, degree of turbulence, and ignition point[6–8]

$$(r_m)(V^{1/3}) = \text{const} = \frac{K_{st}}{10} \text{ or } \frac{K_G}{10} \tag{4-5}$$

Thus, for example, r_m in a 64-m³ sphere is one-fourth the corresponding value in a 1-m³ sphere. (Central ignition gives the highest r_m; if the ignition is not in the center, the flame will contact the close wall before combustion is finished.[4,9] Also, multiple sources of ignition increase dP/dt.[9])

Pressure piling. In compartmented equipment, higher explosion pressures than previously discussed can occur as a result of pressure piling.[10,11] After ignition in one compartment, some of the gas mixture ahead of the flame front is pushed through the connection between the two compartments. Pressure of the original flammable mixture in the second compartment increases, and the resulting compressed mixture is ignited by the flame from the first compartment. The final explosion pressure is related particularly to the size of the connection between the two compartments. Tests have been performed by the U.S. Bureau of

TABLE 4-1 Effect of Initial Pressure on Maximum Explosion Pressure (5% Propane in Air)[5]

Initial pressure (P_i), kPa abs	Maximum explosion pressure (P_m), kPa abs	P_m/P_i
200	1620	8.10
400	3309	8.27
600	4895	8.16

Mines in a closed box, 0.3 by 0.3 m and 1.27 m long (0.11 m^3).[10] Movable 0.3- by 0.3- by 0.013-m partitions with single holes of varying size were used to form compartment ratios of 1:1, 3:1, and 7:1. Both spaces contained 9.5% v/v natural gas ($C_{st} \approx 9.5\%$ v/v). Maximum explosion pressure, illustrating the effect of pressure piling, is shown in Fig. 4-4. There is a semilogarithmic relationship between maximum explosion pressure and hole size:

$$\log p_m = -0.002416 D_h + 0.424208 \qquad D_h = 50.8 \text{ to } 254 \quad (4\text{-}6)$$

The compartment ratio giving the maximum explosion pressure was 1:1, but it was 3:1 for the 203-mm-diameter hole, where ignition was in the larger compartment. Highest pressures with the 203-mm-diameter and smaller holes occurred in the compartment adjoining the ignition compartment.

Turbulence

Maximum explosion pressures P_m are increased only slightly (about 6 percent) at the stoichiometric concentration in air by initial turbulence.[4,6,9] The increase in P_m by initial turbulence is more marked at the lower

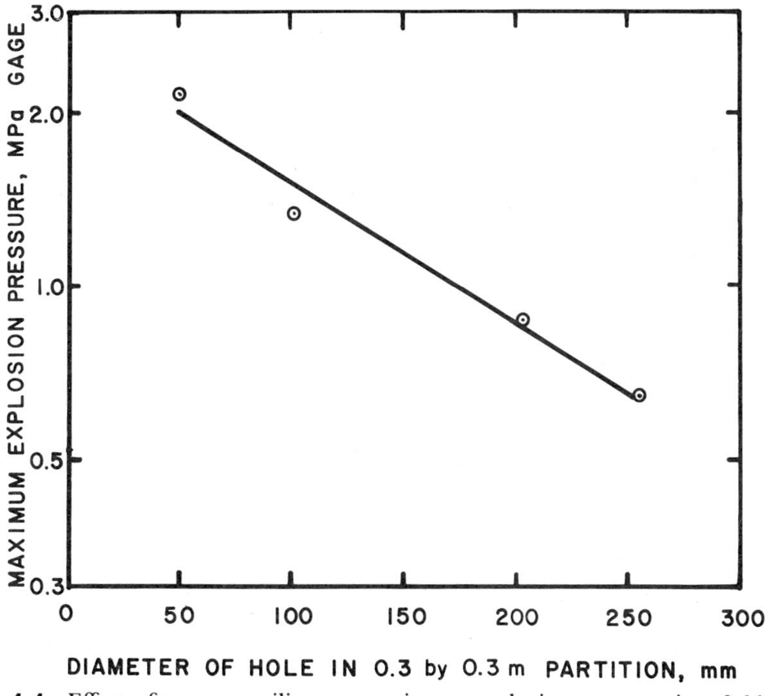

Fig. 4-4 Effect of pressure piling on maximum explosion pressure in a 0.11-m^3 compartmented box with 9.5% v/v natural gas in air. Two compartments separated by a 0.3- by 0.3-m partition. (*Data of Gleim and March.*[10])

and upper flammability limits[4,6]; in tests with methane-air mixtures near the lower flammability limit, P_m was 30 percent more with initial turbulence than without it.[4]

Initial turbulence greatly increases the rates of explosion-pressure rise.[4,6,9] Figure 4-5 shows the effect of turbulence on r_m for 3% pentane in air in a 1.7-m^3 vessel equipped with a 460-mm-diameter fan to provide turbulence (C_{st} = 2.56% v/v). The maximum rate of explosion-pressure rise increases linearly with increase in fan speed; at 2000 r/min r_m is about 5 times the corresponding value without turbulence. Also, with 9.4% v/v methane in air, r_m is about 5 times more with initial turbulence than with no turbulence.[4] Thus, r_m can be at least 5 times more with high initial turbulence than with quiescent conditions.

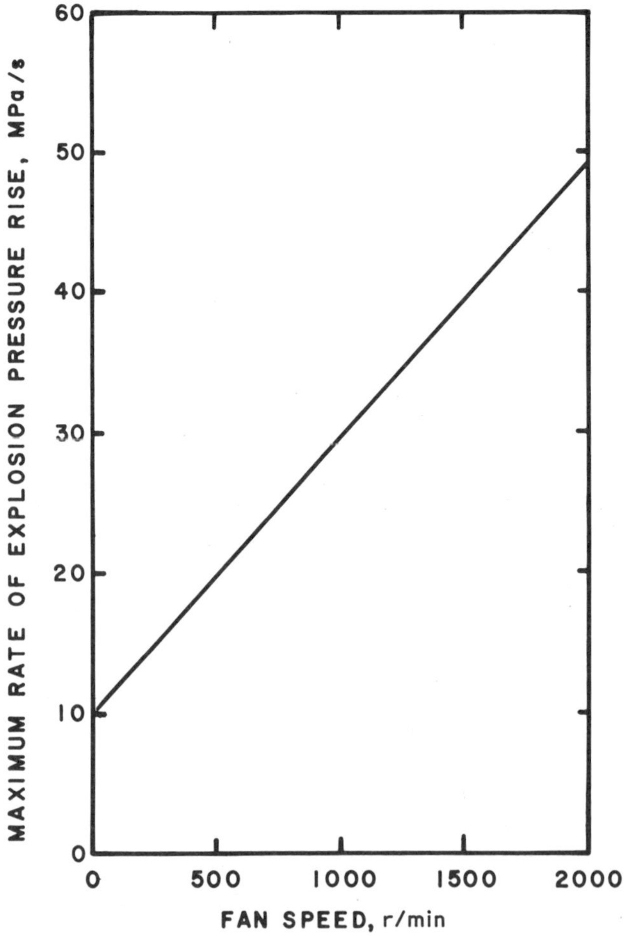

Fig. 4-5 Result of turbulence on maximum rate of explosion-pressure rise in a closed 1.7-m^3 vessel with 3.0% v/v pentane in air. *(After Harris,[6] by permission.)*

Ignition Source

Tests on the effect of the ignition source on P_m and dP/dt have been performed by the Bureau of Mines.[4] The experiments were performed in a 28-L vessel containing 9.4 or 9.5% v/v methane in air using the following igniters:

1. Spark
2. 100 mg guncotton (unpulped nitrocellulose)
3. 250 mg guncotton (unpulped nitrocellulose)
4. 400 mg flash powder (magnesium and sodium perchlorate)

The P_m increases only slightly as the strength of the igniter increases, as shown in Fig. 4-6.

The maximum rate of explosion-pressure rise is the same for the first three igniters in Fig. 4-6. Higher r_m occurs only with the strongest igniter, 400 mg of flash powder. On the other hand, the average dP/dt increases

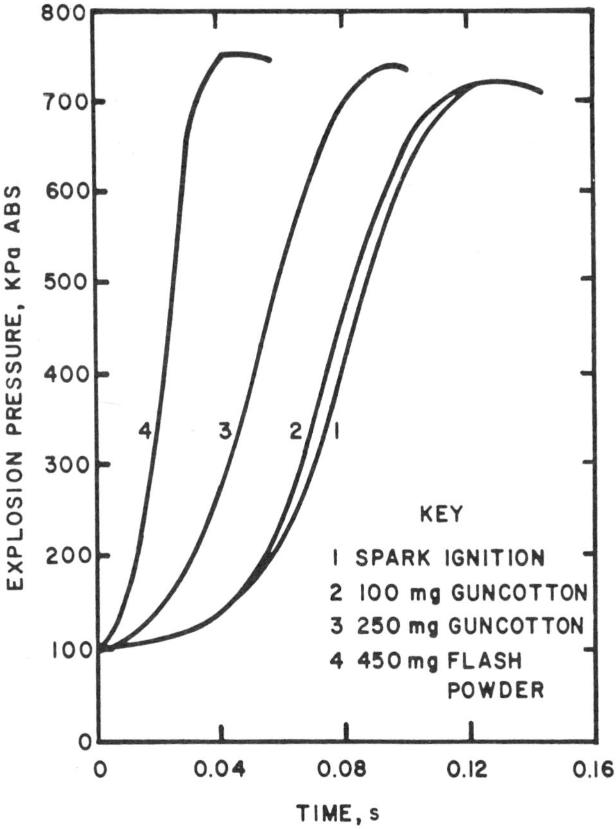

Fig. 4-6 Effect of ignition source on explosion-pressure development in a closed 28-L cubical vessel with 9.4% v/v methane in air. (*After Nagy et al.*[4])

markedly with the 250-mg guncotton and 400-mg flash-powder igniters. Thus, with these ignition sources, the duration of the incipient explosion is decreased greatly. The investigators at the bureau attributed the higher rate of pressure rise to an increase in flame-front area, i.e., akin to the rise produced by multiple sources of ignition, noted in the discussion above of vessel volume and shape. Explosion-pressure rises for the above conditions are summarized in Table 4-2.

Rapid explosions are more difficult to vent than mild explosions. Although more rapid explosions occur with very strong igniters, it is not likely that such ignition sources will occur in industrial plants.

4-4 Transition to Detonation

A deflagration can develop into a gaseous detonation under certain conditions. The range of detonability is narrower than the range of flammability. For example, the range of detonability of hydrogen in air is 18 to 59 percent v/v, compared with a range of flammability of 4 to 75 percent v/v. If the length-to-diameter ratio of a pipe or vessel is more than somewhere around 10 (run-up distance) with flammable gases in air at atmospheric pressure, a detonation is possible. Also, pipe diameter must be above a critical diameter, 12 to 25 mm. (Ignition of natural gas or propane at stoichiometric concentrations in air does not result in detonations in long 25-mm-diameter flame-front tubes to flares. Natural gas has detonated in long 100-mm-diameter pipes, however.[12]) Treatment of the complex theory of gas-phase detonations is beyond the scope of this volume but is covered in Refs. 11 to 16; a description of detonations with flammable gas in air follows.

Detonation Pressure

With ignition at one end of a closed tube, a series of pressure waves traveling at the speed of sound moves through the unburned gas. Later waves traveling through the unburned gas that has been heated by compression from the earlier waves speed up because of the higher temperature and overtake the first wave, and a shock wave develops. Flame fol-

TABLE 4-2 Effect of Ignition Source on Explosion-Pressure Rise in a 28-L Vessel (9.4% Methane in Air at 101.325 kPa abs)[4]

Ignition source	r_m, MPa/s	r_{av},* MPa/s	Time to 200 kPa abs, ms
Spark	11.8	4.65	53
Guncotton, 100 mg	11.8	4.65	51
250 mg	11.8	6.32	28
400 mg flash powder	29.2	14.0	12

*$r_{av} = (P_m - P_i) \times 10^{-3}$ divided by the time in seconds from ignition to P_m.

lows the shock wave and catches up with it, forming a detonation wave. This wave may be momentarily unstable, with pressures and speed in excess of the stable detonation wave that develops from the unstable one. Peak pressure p_{so} in the unstable phase of the detonation wave is about 100 times the initial absolute pressure, as shown at 15 m in Fig. 4-7.

The stable detonation wave is called the *Chapman-Jouquet wave.* It moves with supersonic speed relative to the unburned mixture, and peak pressures are of the order of 30 times the initial absolute pressure, neglecting the momentary higher and normally undamaging pressure that occurs in the von Neumann pressure spike at the leading edge of the detonation wave. Measured pressures in a long 100-mm-diameter pipe are shown in Fig. 4-7.

The stresses in the pipe at the peak pressure are most likely above the ultimate strength of the pipe. Nevertheless, the duration of the unstable detonation is so brief that there is not enough time for much strain to occur, thanks to the inertia of the pipe mass.[12] Thus, the peak pressure is generally too short to be felt by usual process equipment. (Material more than about 25 mm thick, however, may fail in a brittle manner at the peak pressure of an unstable detonation.) The lower pressure occurring in the Chapman-Jouquet plane of the stable detonation wave may be of sufficient duration, albeit short, to damage process piping and equipment. A fissure in a pipeline will slow a detonation wave, but the wave can regain speed and pressure as it travels down a long pipeline.

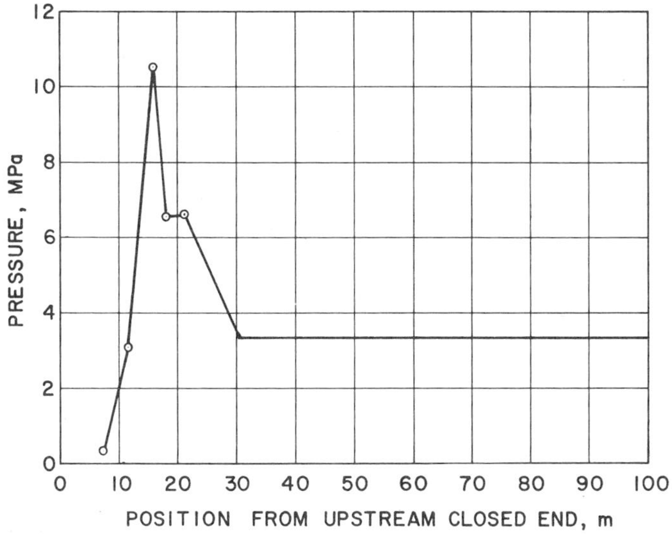

Fig. 4-7 Pressure developed from detonation of a 9.2% v/v natural gas mixture in air (C_{st} = 9%) initially at 0.1 MPa abs. Pipe length = 103 m; pipe diameter = 100 mm. Turbulent flow initially at 1.26 to 1.37 m/s. Pipe ends closed except for an opening to control flow. Ignition by spark plug at upstream closed end. *(After Henderson,[12] by permission.)*

Reflected pressure. The previous discussion pertains to the incident (side-on) p_{so} overpressure. A reflected pressure p_r is developed instantly on a rigid surface if the shock wave impinges on the surface at an angle to the direction of the propagation of the wave. Reflection increases the pressure on the surface above the incident pressure. The reflected pressure is a function of pressure in the incident wave and the angle between the surface and the shock front. The maximum ratio of p_r/p_{so} when a strong shock wave strikes a flat surface head on is 8. Thus, the stable detonation wave may cause enormously high pressures at closed ends of pipes, bends, and tees. For example, with an incident overpressure of 3.5 MPa, a shock wave can develop (3.5)(8) = 28 MPa overpressure on the flat end of a pipe. Furthermore, acceleration from a suddenly applied force of the detonation wave can double the load that a structure feels, producing 56 MPa overpressure in the example above. Also, enhancement of pressure effects can occur at weak points of the metal. Consequently, it is often at elbows, tees, and closed ends of pipes that the greatest destruction from a gaseous detonation occurs.

Pressure piling. As already described in this chapter, relatively high explosion pressure can occur in compartmented equipment from pressure piling. Thus, a restriction in a pipeline, such as an orifice, may result in pressure piling; if a detonation develops in a precompressed volume, greatly augmented detonation pressures can result.

Prevention and Protection

Ordinarily, the prudent course for guarding against the destructive effects of detonations is to prevent the formation of flammable mixtures to the greatest extent practicable, as described in Chap. 2. (Acetylene may detonate in the absence of air. Protection of equipment against detonation in acetylene systems has been described by Sargent[17] and Miller and Penny.[18]) Nevertheless, some measures can be taken to reduce, if not eliminate, the destructive effect of a detonation should one occur unexpectedly.

Strength of equipment. No inviolable rule can be given on design pressure to protect against detonations. Equipment designed to contain a pressure of 3.5 MPa, however, usually will be adequate to contain a detonation, with other safeguards, discussed in the following sections, for flammable gases in air at atmospheric pressure.

Geometry. Large length-to-diameter ratios promote the development of detonations; vessels should be designed with the lowest feasible length-to-diameter ratio if a detonation is possible. Dished heads on vessels survive detonations better than flat heads because of the more unfavorable angle of incidence with the latter. If turns in a process line are required, two 45° bends will greatly reduce reflected pressure compared with a single 90° elbow.

Restrictions in pipelines, such as orifices, may intensify a detonation by promoting pressure piling. (In some cases, restrictions may block a detonation, but this possible amelioration cannot be relied upon because of the threat of intensification.) On the other hand, a detonation is extinguished when it enters a wider pipe from a smaller one. Nevertheless, it may be regenerated somewhere along the larger pipe. Cubbage[19] has described experiments which showed that a detonation degenerated on passing from a 25- to a 100-mm-diameter pipe. After traveling about 2.4 m in the wider pipe, however, the flame speed increased and a stable detonation was reestablished after an additional 4.3 to 5.2 m. Thus, enlargements in pipe diameter at strategic locations, such as entrances to elbows and tees, can act to squelch a detonation temporarily and thereby prevent damage at susceptible locations.

Flame arresters. Cubbage[19] also found that a flame arrester of crimped metal ribbon with a diameter 4 times that of the pipe can arrest a detonation with town gas in air and with gases having comparable flame speed. (Town gas is approximately 3% v/v CO_2, 1.4% unsaturated hydrocarbons, 0.6% O_2, 17% CO, 14.5% CH_4, 2% higher paraffins, 52% H_2, and 9.5% N_2.) The crimped ribbon consists of two strips of metal, one crimped and the other flat. Flow is through the triangular opening in the crimped piece.* These flame traps reportedly are effective in damping detonations at any position in pipelines; it is preferable to locate them as near as possible to the most probable source of ignition, such as burners. Furthermore, additional traps can be installed every 10 m or so upstream of the former one if additional ignition sources are possible. Typical pressure drops are shown in Table 4-3.

Proper sealing of the arrester in the housing is essential to stop flame.

TABLE 4-3 Pressure Drops for Crimped-Metal-Ribbon Flame Arresters (Data from Ref. 19)

Rate of flow, m³/h	Pipe diameter, mm	Pressure drop, Pa
30	50	62
30	75	25
30	100	8
150	75	210
150	100	100
150	150	45
300	100	330
300	150	135

*This arrester is supplied by Amal, Ltd., Birmingham, England, B6 7ES.

Also, a flame arrester can overheat and transmit flame if flame stabilizes on the arrester; remedies to counter overheating are described in Chap. 6.

Packed-tower flame arresters to arrest deflagration or detonation in acetylene transmission systems are described in Pamphlet G1.3 of the Compressed Gas Association. They should also be capable of arresting detonations of flammable gases in air.

Rupture disks. The advantage of rupture disks in moderating the damaging effects of detonations is moot. Nevertheless, rupture disks can prevent precompression of gas in compartmented equipment and thereby decrease the potential for destructive detonative pressures resulting from pressure piling. Also, rupture disks at closed ends of pipes, bends, and tees can be used to prevent damage at these locations from reflected pressure; protection from a probable fire at the outlet requires due consideration. Otherwise, no firm guidance on combating detonations by use of rupture disks can be given; the necessary liberal use of explosion venting to prevent a transition to detonation would be impracticable.

4-5 Blast Effects

Vessels, barricades, or rooms may contain detonable solids or liquids, and it may be necessary to assess the consequences of a detonation on the structure. Also, the external effects of a blast on adjacent equipment often requires review. Detailed analysis of these special phenomena, such as air-blast loading plus fragmentation and missiles, is beyond the scope of volume. Such an analysis will be found in Ref. 20 to 24; some rudimentary considerations on blast effects, however, are covered in the following sections.

Energy

Bursting vessel. The maximum energy in a bursting vessel is that released by the isentropic expansion of the gas in the vessel from the burst pressure P_b to the pressure of the surrounding gas, usually air, P_s (MPa)[21]

$$E = \frac{P_b V}{\kappa - 1}\left[1 - \left(\frac{P_s}{P_b}\right)^{(\kappa - 1)/\kappa}\right] \qquad (4\text{-}7)$$

The volume V in Eq. (4-7) is for the gas phase; κ is for the gases existing at rupture. With atmospheric pressure of 0.1 MPa, the energy E from rupture of a 4-m³ cylinder of air ($\kappa = 1.4$) at an absolute pressure of 1.5 MPa is 8.1 MJ. With isobutane ($\kappa = 1.11$) for the same conditions the blast energy is 12.8 MJ. Ordinarily, the peak pressure of the shock will be equivalent to the bursting pressure of the vessel, which will decay according to the scaling laws discussed later in this chapter. The correction in Eq. (4-7) for kinetic energy imparted to missiles is insignificant. Also, the amount of flammable liquid is unimportant except to the extent that it reduces the volume of the gas phase in the vessel. In case of rupture, however, the secondary fire or explosion can be disastrous, particularly for releases of flammable materials inside buildings; in these circumstances, the amount of material can be very important (see the discussion of sight glasses and flexible hoses in Sec. 7-3). A vapor-cloud explosion, described in Chap. 6, could result from gross emission of flammable gas following rupture. More likely than not, though, with a probable ignition source from tearing metal, just a fire would occur. High[25] determined that the diameter of fireballs D_f, in meters, from many explosions with W_f kg of flammable material can be estimated by

$$D_f = 3.86 W_f^{\,0.32} \tag{4-8}$$

The blast energy resulting from failure of a vessel from a gas-phase deflagration falls in the bursting-vessel category.

Internal detonations. Loving[26] developed Eq. (4-9) for estimation of equivalent hydrostatic pressure from detonation of condensed explosives in rooms. Charges up to 4.5 kg were used in the tests. The pressure estimate is for firing unconfined charges in the center of a room with the maximum dimension no more than twice the minimum dimension; detonations close to a wall will develop more pressure than estimated by

$$p_m = \frac{k'W}{V} \tag{4-9}$$

The constant k' for TNT, PETN (pentaerythritol tetranitrate), and 40% dynamite is 8.6, 6.4, and 3.0, respectively. Thus, with 4 kg of TNT or its equivalent in a 30-m³ vessel, $p_m = 1.15$ MPa gage.

If the vessel fails, the type of confining structure does not have a major effect on the blast effects from detonation of condensed materials. In this case,

$$E = (W)(-\Delta H) \tag{4-10}$$

where ΔH is in megajoules per kilogram. (Hazardous compounds and reactions are covered at greater length in Chap. 7.) Deflagration of condensed hazardous compounds is more difficult to evaluate. Nevertheless, when the mass of material per unit volume (loading density) is high, it is generally judicious to assume that the blast-energy potential is approximately the same as from a detonation in Eq. (4-10).

The energies determined from Eqs. (4-7) and (4-10) are for airburst situations and should be doubled if the blast occurs on or near the ground; an explosion may be considered an air burst if its center is at such a height that spherical expansion of the wave to the point of interest will not be affected by reflections. The TNT equivalent weight is obtained by dividing E from Eqs. (4-7) and (4-10) by the specific energy of TNT, 4.52 MJ/kg. (In determining the TNT equivalent weight the multiplication factor of 2 for surface bursts is not needed.) Thus, an E of 500 MJ is equivalent to 110 kg TNT. Analogously, for the examples of bursting 4-m^3 vessels covered in the preceding section, the TNT equivalent weights for air and isobutane are 1.8 and 2.8 kg, respectively.

Blast Pressure

When energy is suddenly released to the ambient air, a shock wave develops; initially it travels through the air at supersonic speed, but as the intensity of the wave subsides, it becomes sonic. A shock wave in air is usually referred to as a *blast wave* because it may be accompanied by a strong wind; the front of the blast wave is the shock front. A blast produces various effects, depending on its intensity. Only blasts originating on the ground will be discussed, and primary emphasis is on side-on blast overpressures p_{so} below about 200 kPa.

When a blast wave impacts on an object, the pressure rises essentially instantaneously to a peak incident (side-on) pressure p_{so} on surfaces oriented parallel to the direction of the wave. If the shock front strikes a solid surface at an angle, reflection occurs. The reflected pressure on the surface depends on p_{so} and the angle of incidence. (The angle of incidence is the angle between the shock front and the reflecting surface. An angle of incidence of 90° is for a wave moving parallel to the surface; an angle of 0° means that the blast wave strikes the surface head on.) For weak shocks, $p_r = 2p_{so}$. The p_r increases up to 8 p_{so} for strong shocks making a direct strike on an object, as indicated by[22]

$$\frac{p_r}{p_{so}} = 2\,\frac{7P_s + 4p_{so}}{7P_s + P_{so}} \tag{4-11}$$

where P_s is in kilopascals.

The ratio p_r/p_{so} stays close to that determined from Eq. (4-11) up to angle of incidence of about 40°, whereafter it drops to one with an angle of incidence of 90°. In addition to overpressure from the shock front, a structure is also subjected simultaneously to dynamic pressure, i.e., wind, in a blast wave. The peak wind velocity behind the shock front depends on the peak overpressure behind the shock front, as shown in Table 4-4.

The changes of overpressure and dynamic pressure with time at a fixed point are shown schematically in Fig. 4-8. The negative amplitude of the pressure, as shown in Fig. 4-8, is much smaller than peak positive incident pressure, and in any event the peak negative incident and negative normal reflected pressures do not exceed 100 kPa gage. (The duration of the

TABLE 4-4 Peak Overpressure, Wind Velocity, and Dynamic Pressure in Air at Sea Level for an Ideal Shock Front (Data from Ref. 22)

Peak overpressure, p_{so}, kPa, gage	Wind speed, m/s	Dynamic pressure, kPa, gage
175	275	90
125	210	50
75	145	20
25	50	2
15	30	0.7

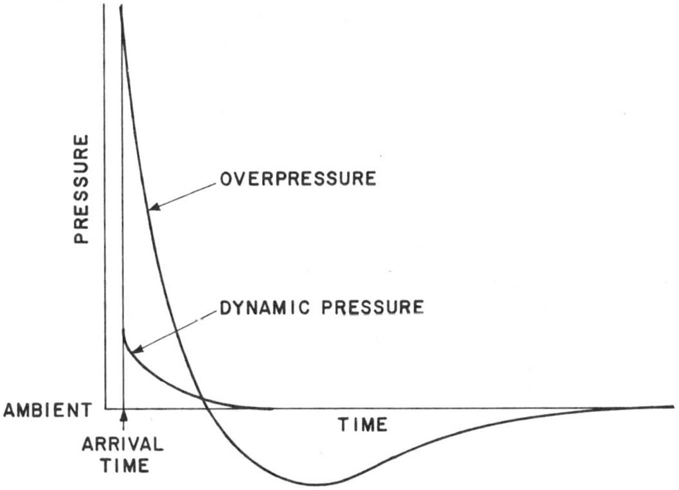

Fig. 4-8 Variation of incident overpressure and dynamic pressure (wind) with time at a fixed location. (*From Glasstone.*[22])

negative pressure is approximately $10W^{1/3}$ms for explosions at the surface.[20]) Even so, structures that withstand positive overpressures may not survive lower negative pressures because of less support on the outside than on the inside of buildings.

Scaling laws. The shock-wave parameters for a given substance are equivalent[27,28] for the same value of scaled distance, $Z = R/W^{1/3}$. Consequently, for example, peak overpressure will be the same at R_1 m from a blast of W_{TNT_1} kg as at R_2 m from W_{TNT_2} kg, according to the relationship

$$\frac{R_1}{R_2} = \left(\frac{W_{TNT_1}}{W_{TNT_2}}\right)^{1/3} \tag{4-12}$$

Positive overpressures, scaled positive pressure impulses $i_s/W_{TNT}^{1/3}$ and $i_r/W_{TNT}^{1/3}$, and scaled durations of the positive pressure phase $t_0/W_{TNT}^{1/3}$ for hemispherical TNT surface blasts at sea level are shown in Fig. 4-9. The same parameters for a spherical TNT detonation in air at sea level are shown in Fig. 4-10. (It is not necessary to alter W_{TNT} in

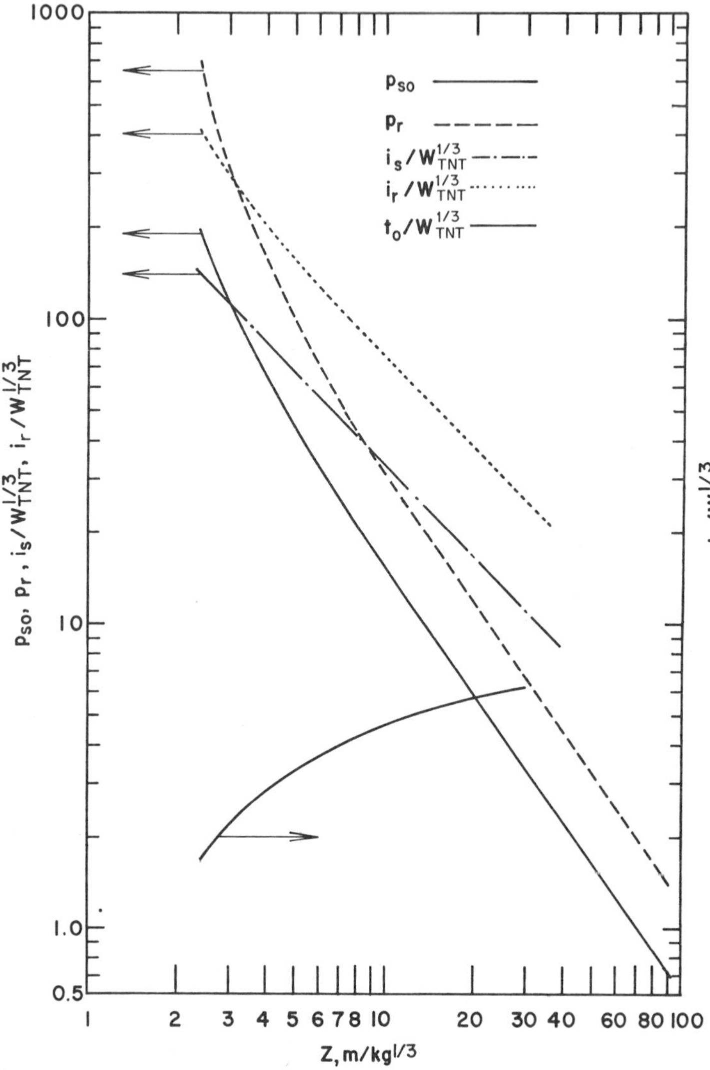

Fig. 4-9 Shock-wave parameters for hemispherical TNT surface explosion at sea level. *(After Ref. 20.)*

Z in these figures to account for reflection in surface explosions. Because of reflection in surface explosions, however, overpressures in Fig. 4-9 are $2^{1/3} = 1.26$ times the corresponding pressures in Fig. 4-10 for Z greater than about 10.)

Atmospheric temperature inversions refract a shock wave; below a p_{so} of about 25 kPa, a given overpressure with a surface inversion may be experienced at 2 to 3 times the distance determined from Fig. 4-9.[22,29] On the other hand, a shock wave is refracted upward in an unstable

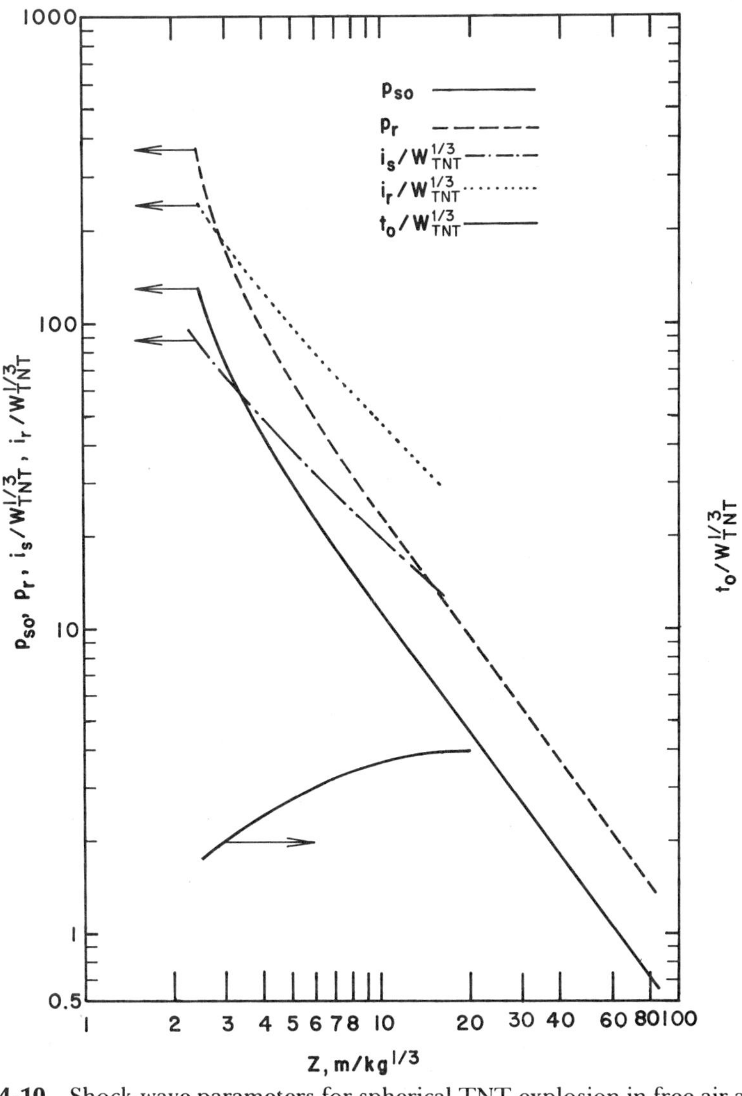

Fig. 4-10 Shock-wave parameters for spherical TNT explosion in free air at sea level. *(After Dobbs et al.,[32] by Permission, and Ref. 20.)*

atmosphere, wherein the temperature decreases with height more than 1°C per 100 m. Thus, lower pressures than determined from Fig. 4-9 are likely on a hot summer day when atmospheric instability occurs most frequently. Obstacles do not stop an air blast; it diffracts around the obstacle and reforms at essentially full strength within five obstacle dimensions beyond the obstruction.[23]

Response to overpressure. Some pressure criteria for stated damage to structures are listed in Table 4-5. In addition, 50 percent glass breakage can occur at 0.7 kPa gage pressure.[30] Both the peak overpressure and the peak dynamic (wind) pressure determine the amount of damage, but for certain structures one or the other of these pressures has the dominant effect; damage to land transportation equipment, for example, is due chiefly to wind accompanying a shock front, while items in Table 4-5 are affected primarily by shock. Total destruction is probable above that 70 kPa, and cratering occurs above approximately 1930 kPa.[31]

People can tolerate fairly high pressures without injury. Thus, the thresholds of fast-rise overpressures of short duration (3 to 5 ms) for lethality, lung damage, and eardrum rupture are 690 to 1380, 207 to 255, and 34 kPa, respectively; 50 percent of the people suffer eardrum rupture at 100 kPa.[20] These physiological effects are for incident pressure, the incident plus the dynamic pressures, or the reflected pressure. Nevertheless, injuries or fatalities can occur at lower blast overpressures from missiles or flying debris and collapse of structures on personnel. Also, a person can be knocked down at an incident overpressure of about 15 kPa. Thus, impact of a body on a hard surface can also cause an injury or a fatality.

TABLE 4-5 Damage From Overpressure (Data from Ref. 22)

Structure	Damage*	Approximate incident overpressure for stated damage, kPa gage†
Corrugated asbestos siding	Shattering	7–14
Corrugated steel or aluminum paneling	Connection failure followed by buckling	7–14
Wood siding panels, standard house construction	Usually failure occurs at the main connections, allowing a whole panel to be blown in	7–14
Concrete or cinderblock wall panels, 200 or 300 mm thick (not reinforced)	Shattering of the wall	14–21
Brick wall panel, 200 or 300 mm thick (not reinforced)	Shearing and flexure failures	48–55

*There is only a small difference between the overpressures causing little or no damage and complete failure for these structures.

†Incident (side-on) overpressures are for panels that face the explosion. If there is no reflected pressure, the indicated pressures must be doubled.

Example

Distances from the blasts of the 4-m³ air cylinder (1.8 kg TNT) and the 110-kg TNT explosion at the ground, cited in the energy portion of Sec. 4-5, are given in Table 4-6. They are based on Fig. 4-9 and Table 4-5.

Often the lack of damage to a structure from a nearby blast can also be used to ascertain the intensity of the blast. Thus, assume that a 200-mm-thick unreinforced concrete wall facing an explosion 100 m away is demolished ($p_{so} = 14$ and $Z = 10.6$) but the adjoining walls with no reflected pressure are only cracked. From Table 4-5 and Fig. 4-9 $p_{so} \approx 28$ and $Z \approx 6.8$ for these adjoining sides. Accordingly, $W_{TNT} \approx 840$ to 3180 kg, and other observed effects may make it possible to estimate the energy of the explosion more precisely.

TABLE 4-6

Effect	Scaled distance, Z, m/kg$^{1/3}$	Distance R, m: From air cylinder	Distance R, m: From 110-kg TNT explosion
Probable total destruction ($p_{so} = 70$)	3.8	4.6	18.2
Eardrum rupture ($p_{so} = 34$)	5.9	7.2	28.3
Personnel knocked down ($p_{so} = 15$)	10.1	12.3	48.4
Shattering of concrete wall, 200 mm thick, facing blast and not reinforced ($p_{so} = 14$)	10.6	12.9	50.8
50% glass breakage ($p_{so} = 0.7$)	87.3	106.2	418.3

References

1. Zabetakis, M. G., "Flammability Characteristics of Combustible Gases and Vapors," *U.S. Bur. Mines Bull.* 627 (*USNTIS* AD 701 576), 1965.

2. National Fire Protection Association, Guide for Explosion Venting, *NFPA* 68, Boston, 1974.

3. McKinnon, G. P. (ed.), *Fire Protection Handbook,* 14th ed., sec. 2, chap. 2, National Fire Protection Association, Boston, 1976.

4. Nagy, J., E. C. Seiler, J. W. Conn, and H. C. Verakis, "Explosion Development in Closed Vessels," *U.S. Bur. Mines Rep. Invest.* 7507, April 1971.

5. Cousins, E. W., and P. E. Cotton, "Design Closed Vessels to Withstand Internal Explosions," *Chem. Eng.,* vol. 58, no. 8, pp. 133–136, August 1951.

6. Harris, G. F. P., "The Effect of Vessel Size and Degree of Turbulence on Gas Phase Explosion Pressures in Closed Vessels," *Combust. Flame,* vol. 11, pp. 17–25, February 1967.

7. Bartknecht, W., "Report on Investigations on the Problem of Pressure Relief of Explosions of Combustible Dusts in Vessels," *Staub Reinhalt. Luft,* vol. 34, no. 11, pp. 289–300, November 1974.

8. Zabetakis, M. G., "Fire and Explosion Hazards at Temperature and Pressure

Extremes," *AIChE–Inst. Chem. Eng. Symp. Ser.* 2, *Chem. Eng. Extreme Cond. Proc. Symp., 1965,* pp. 99–104.

9. Maisey, H. R., "Gaseous and Dust Explosion Venting, Part 1," *Chem. Process Eng. (Lond.),* vol. 46, no. 10, pp. 527–535 and 563, October 1965.

10. Gleim, E. J., and J. F. March, "A Study to Determine Factors Causing Pressure Piling in Testing Explosion-Proof Enclosures," *U.S. Bur. Mines Rep. Invest.* 4904, August 1952.

11. Munday, G., "Detonations in Vessels and Pipelines," *Chem. Eng. (Lond.),* no. 248, pp. 135–144, April 1971.

12. Henderson, E., "Combustible Gas Mixtures in Pipe Lines," *Proc. Pac. Coast Gas Assoc.,* vol. 32, pp. 98–111, September 1941.

13. Jost, W., *Explosion and Combustion Processes in Gases,* McGraw-Hill, New York, 1946.

14. Lewis, B., and G. von Elbe, *Combustion, Flames, and Explosions of Gases,* 2d ed. Academic, New York, 1961.

15. Penner, S. S., and B. P. Mullins, *Explosions, Detonations, Flammability and Ignition,* Pergamon, New York, 1959.

16. Stull, D. R., Fundamentals of Fire and Explosion, *AIChE Monogr. Ser.,* vol. 73, no. 10, 1977.

17. Sargent, H. B., "How to Design a Hazard-Free System to Handle Acetylene," *Chem. Eng.,* vol. 64, no. 2, pp. 250–254, February 1957.

18. Miller, S. A., and E. Penny, "Hazards in Handling Acetylene in Chemical Processes Particularly under Pressure," *Inst. Chem. Eng. Symp. Ser.* 7, *Proc. Symp. Chem. Process Hazards Spec. Ref. Plant Des., 1960,* pp. 87–94.

19. Cubbage, P. A., "The Protection by Flame Traps of Pipes Conveying Combustible Mixtures," *Inst. Chem. Eng. Symp. Ser.* 15, *Proc. 2d Symp. Chem. Process Hazards Spec. Ref. Plant Des., 1963,* pp. 29–34.

20. Anon., Structures to Resist the Effects of Accidental Explosions, *Dept. Army Tech. Man.* TM5-1300, *Dept. Navy Publ.* NAVFACP-397, *Dept. Air Force Man.* AFM88-22, June 1969.

21. Baker, W. E., *Explosions in Air,* University of Texas Press, Austin, 1973.

22. Glasstone, S., *The Effects of Nuclear Weapons,* rev. ed., U.S. Atomic Energy Commission, Washington, 1962.

23. Lawrence, W. E., and E. E. Johnson, "Design for Limiting Explosion Damage," *Chem. Eng.,* vol. 81, no. 1, pp. 96–104, Jan. 7, 1974.

24. Moore, C. V., "The Design of Barricades for Hazardous Pressure Systems," *Nucl. Eng. Des.,* vol. 5, no 1 pp. 81–97, January–February 1967.

25. High, R. W., "The Saturn Fireball," *Ann. N.Y. Acad. Sci.,* vol. 152, art. 1, pp. 441–451, 1968.

26. Loving, F. A., "Barricading Hazardous Reactions," *Ind. Eng. Chem.,* vol. 49, no. 10, pp. 1744–1746, October 1957.

27. Hopkinson, B., *Br. Ord. Board Min.* 13565, 1915.

28. Sachs, R. G., The Dependence of Blast on Ambient Pressure and Temperature, *Ballistics Res. Lab. Rep.* 466, Aberdeen, Md., 1944.

29. Brasie, W. C., and D. W. Simpson, "Guidelines for Estimating Damage Explosion," *Chem. Eng. Prog. 2d Loss Prev. Symp., St. Louis, 1968,* pp. 91–102.

30. Strehlow, R. A., and W. E. Baker, "The Characterization and Evaluation of Accidental Explosions," *Prog. Energy Combust. Sci.,* vol. 2, no. 1, pp. 27–60, 1976.

31. Robinson, C. S., *Explosions: Their Anatomy and Destructiveness,* McGraw-Hill, New York, 1944.

32. Dobbs, N., E. Cohen, and S. Weissman, "Blast Pressures and Impulse Loads for Use in the Design and Analysis of Explosive Storage and Manufacturing Facilities," *Ann. N.Y. Acad. Sci.,* vol. 152, art. 1, pp. 317–338, 1968.

5

Explosion Protection

Methods to prevent explosions in equipment were examined in Chap. 2. It is generally not essential to back up adequate explosion-prevention methods with explosion-protection facilities. Sometimes, however, prevention of an explosion may be impractical, although the chance of one's occurring may be low. In such cases, explosion-protection equipment must be provided. Also, in some instances, depending on the risk and consequences of an explosion, protection facilities may be desirable in conjunction with prevention measures. Thus, explosion protection by containment, explosion suppression, and explosion venting is discussed in the following sections of this chapter. (Protection from unconfined vapor-cloud explosions is discussed in Chap. 6. Protection from gaseous detonations was considered in Chap. 4. Additional aspects of protection from dust explosions are covered in Chap. 8.)

5-1 Containment

Equipment may be designed to withstand without damage the maximum explosion pressure developed by the material being processed. The special engineering technology to accomplish this is beyond the scope of this book. [The American Society of Mechanical Engineers (ASME) Boiler and Pressure Vessel Code, section VIII, division 1, covers the design, fabrication, and inspection of pressure vessels. Chuss[1] has prepared a simplified version of the code.] Containment should always be considered. Duxbury[2] indicates that costly research effort may be required to obtain the necessary data on explosion venting for polymerization reactors, whereas calculation of the peak pressure in the absence of vents is much simpler. Also, possible problems of flame plus the need for proper collection and disposal of the discharged material from an explosion vent often make containment an attractive protection method. Furthermore, some substances might be too toxic for venting, and containment is necessary in such cases. The ASME code does not offer guidance on explosions, and the pressure to be selected for containment often is moot. The ASME test pressure is sometimes used as the containment pressure. On the other hand, excursion to a higher pressure may be tolerable if only yielding occurs, for example, without extensive equipment damage or injury, considering the usually low probability of an explosion; the Factory Mutual Engineering Corporation in its Loss Prevention Data Sheet 7-76, Combustible Dusts, uses 90 percent of the ultimate strength of the metal for design purposes. As an additional guide, *NFPA* 85F, Standard for the Installation and Operation of Pulverized Fuel Systems, specifies that pulverized-fuel–air systems are to be designed to withstand an explosion pressure of 345 kPa gage for containment of possible internal explosions. (The maximum explosion pressures of pulverized fuels are about 690 kPa gage, i.e., about twice the 345-kPa gage containment pressure.)

Whatever containment pressure is used, appendages on the vessel must also be able to withstand the pressure so as not to be a weak link. An explosion can propagate against normal gas flow when an incipient explosion in a vessel produces more pressure than a blower develops. Thus, connecting vessels may all require the necessary strength for containment. As noted in Chap. 7, distillation columns may be fabricated to contain an explosion, yet extensive internal damage can occur to plates even though an explosion is successfully contained otherwise. Explosion prevention in such cases is the prudent course. Furthermore, containment of an explosion may not always be practical, and in those cases feasible alternatives for explosion control must be adopted.

5-2 Explosion Suppression

During the incipient stage of an explosion, the pressure rises relatively slowly, as shown in Fig. 4-2. In an explosion-suppression system, pressure of only about 3.5 kPa gage in this incipient period is sensed by a pressure-sensitive diaphragm detector.[3–6] Ultraviolet detectors are also used. The diaphragm movement closes an electric circuit to start suppression; an explosive actuator breaks a disk in the outlet of extinguishers to release the suppressant instantly by high-pressure nitrogen to squelch a vapor, gas, or dust explosion. The suppressants are halogenated hydrocarbons, water, or dry chemical.[3–7] The suppressors come in a variety of sizes and shapes. Some components of an explosion-suppression system are shown in Figs. 5-1 and 5-2.

Successful suppression limits explosion-pressure buildup to about 17 kPa gage (118 kPa abs), as indicated in Fig. 5-3. (The pressure in the

Fig. 5-1 Explosion-suppression-system components. Clockwise from lower right, power unit, high-rate-discharge extinguisher, pressure detector, and ultraviolet detector. *(Fenwal Incorporated.)*

Fig. 5-2 Closeup of explosion-suppression pressure detector. *(Fenwal Incorporated.)*

normal unsuppressed explosion would reach a maximum of about 717 kPa gage.)

Besides suppression of the explosion, the power unit can activate additional explosion-protection devices for advance inerting, explosion venting, and isolation or shutdown of plant equipment.[3–5]

Explosion suppression is a proved technology and should be considered as a candidate for explosion protection. The NFPA has published a standard on explosion-suppression systems.[6] Manufacturers should be consulted on design, installation, and maintenance.

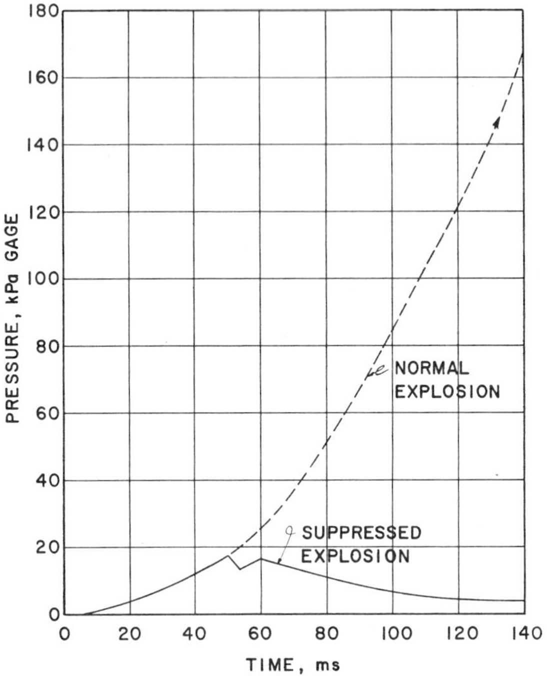

Fig. 5-3 Typical pressure-time curves for normal and suppressed dust explosions. *(After Charney and Lawler,*[3] *by permission.)*

5-3 Explosion Venting

Explosion venting is a third method of explosion protection; ordinarily venting is used to guard against deflagrations, not detonations. The technology of explosion venting is not exact and is still under review. Nevertheless, with proper consideration of the several factors involved and empiricism, venting of explosions can provide adequate protection. Often, however, tests may be necessary to aid in determining the required venting area. Consideration of geometry, such as arrangement of equipment and ducts, is essential to successful explosion venting. The discussion in the following sections applies to both gases and dusts.

Explosion Vent Area

A great deal of research has been performed on explosion venting. Apposite data, information, and reviews are included in Refs. 8 to 41.

The standard burning velocity S_u is a fundamental property of a combustible and is the rate at which an adiabatic plane combustion wave moves relative to the oncoming unburned-fuel–oxidant mixture in a direction perpendicular to the flame surface.[42] The unburned stream is at room temperature and an absolute pressure of 101.325 kPa. Pressure decreases S_u slightly. Temperature has a greater effect,

$$S_u = S_{ui} \left(\frac{T}{T_i} \right)^2 \tag{5-1}$$

where the subscript i refers to the initial standard conditions. Burning velocities are low; the maximum laminar S_u for propane in air is 0.46 m/s. The S_u of most saturated hydrocarbons and solvents is similar to that for propane. (Flame speed is not to be confused with burning velocity. The former is relative to a fixed observer and may be much greater than burning velocity.)

Several investigators have related the rate of explosion-pressure rise dP/dt to S_u. Anthony[8] examined many explosion-venting formulas and concluded that the mathematical approach by Yao[41] using burning velocity is the most promising approach. Nevertheless, he indicated that the major weakness of Yao's work is the need to introduce an arbitrary turbulence factor as a multiplier to the burning velocity. Furthermore, it has been tested only on a small scale. Thus, at this stage in the development of explosion-venting technology, it is prudent to base explosion-venting areas on empirical methods.

Equilibrium venting. In an explosion, the pressure increases due to combustion. Simultaneously, in a vented explosion the pressure tends to decrease due to discharge of gases out the *safety head,* which is a rupture disk and the component parts for holding the disk in place. If the vent is large enough, maximum vented explosion pressure p_v can be kept close to the burst pressure of the disk p_b, that is, equilibrium venting. Appendix B shows the A_v/V ratio at equilibrium venting, based on Ref. 43, to be

$$\frac{A_v}{V} = \frac{7.76r}{Y} \sqrt{\frac{(K)(\mathrm{MW})}{\Delta P \, P_1 T_1}} \tag{5-2}$$

where P_1 in the vessel equals $p_b + 101$ and r is at p_b.

Equation (5-2) implies that smaller explosion vent areas are needed with higher burst pressure of the rupture disk. Nevertheless, r increases greatly with increasing explosion pressure, as indicated in Figs. 4-2 and 5-4. Normally this will overcompensate for the higher venting pressure; for a given maximum vented explosion pressure p_v lower explosion venting area A_v is needed if p_b is close to the operating pressure instead of substantially above it. Consequently, p_b should be set as close to the operating pressure as practicable. (Disk manufacturers require that the operating pressure be held to a specified percentage of the disk rating.) Generally, r is an exponential function of

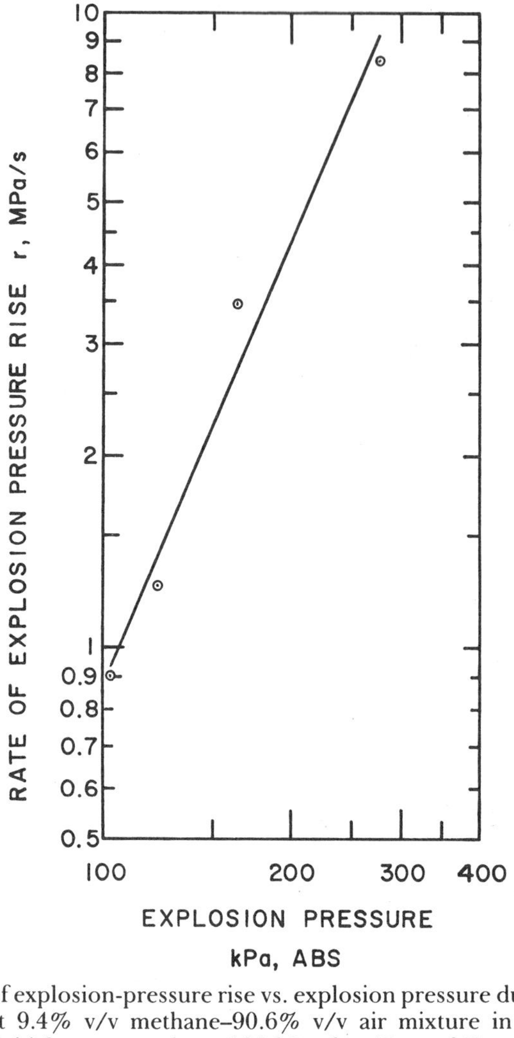

Fig. 5-4 Rate of explosion-pressure rise vs. explosion pressure during explosion of nonturbulent 9.4% v/v methane–90.6% v/v air mixture in a closed 28-L cubical vessel. Initial pressure about 100 kPa abs. (*Data of Nagy et al.* [44])

P early in the explosion process, as shown in Fig. 5-4 and Eq. (5-3) for the methane-air explosion of Fig. 4-2[44]

$$r = bP^{2.32} \tag{5-3}$$

where P is in kilopascals absolute and $b = 20 \times 10^{-6}$ MPa/s·kPa$^{2.32}$.

Example

Assume two venting conditions with the explosions starting at an absolute pressure of 101 kPa and with p_b at 20 and 150 kPa gage. Significant amounts of unburned material may be discharged in the incipient explosion. Therefore, with an initial temperature of 25°C, the minimum temperature at the onset of venting will be from adiabatic compression ($\kappa = 1.4$). With $K = 2$ and $\kappa = 1.4$, Y can be found from Ref. 43. For MW = 29, A_v/V and values of variables in Eq. (5-2) are tabulated below.[r is from Eq. (5-3).]Thus, more

P_1, kPa, abs.	ΔP, kPa, abs.	T_1, K	Y	r, MPa/s	A_v/V, m^{-1}
121	20	314	0.90	1.36	0.10
251	150	388	0.63	7.38	0.18

venting area is required with the higher burst pressure of the rupture disk.

Low-pressure venting. The NFPA[31] has developed a guide for explosion venting for dusts and gases inside vessels based on Richtlinie 3673 of the Verein Deutscher Ingenieure (VDI) in the Federal Republic of Germany.* (The described work is included in appendixes of *NFPA* 68; appendixes are not a part of the NFPA document but are included for information purposes only.) In implementation of the NFPA method, tests are conducted in approximately spherical vessels of about 1 m^3 volume; tests in a vessel of this size are considered more reliable than in the small 1.23-L Hartmann test apparatus (U.S. Bureau of Mines) for extrapolation to plant vessels. From the cubic law in Eq. (4-5) the constants K_{St} for dust and K_G for gases are determined in these tests. (The unit of these K's in the VDI and NFPA publications is bar-meters per second.) Thus, by the cubic law for specified p_v and p_b,[31]

$$\frac{A_{v_2}}{V_2} = \frac{A_{v_1}}{V_1}\left(\frac{V_1}{V_2}\right)^{1/3} \tag{5-4}$$

where subscript 1 is for the test vessel and subscript 2 for the second vessel. The ratio of vessel length to diameter should not be over 5 for application of the cubic law.

Dust hazards are divided into three classes, as listed in Table 5-1.

The strong ignition source can be characterized, for example, as a tongue of burning dust traveling from one vessel to another through a duct.[31] *Smaller explosion venting areas than determined by equilibrium venting*

*Special credit is due to Dr. Ing. W. Bartknecht, Ciba-Geigy AG, Basel, Switzerland, and Ing. C. Donat, Hoechst Limited, Frankfurt-am-Main, West Germany.

TABLE 5-1 Classification of Dust Explosion Hazards[31]

Hazard class	K_{St}, bar·m/s — Weak ignition source (≈10 J)	K_{St}, bar·m/s — Strong ignition source (≈10,000 J)	r_m in 1.23-L Hartmann test apparatus, MPa/s
St-1	≤100	≤200	≤50
St-2	101–200	201–300	50–152
St-3*	>200	>300	>152

*Uncommon.

considerations can be used if overshoot above p_b is permissible. (The ASME Boiler and Pressure Vessel Code does not provide guidance on explosions. The NFPA[31] indicates that p_v should be no more than two-thirds of the pressure which will cause the weakest part of the vented vessel to break. The Factory Mutual Engineering Corporation in its Loss Prevention Data Sheet 7-76, Combustible Dusts, uses 90 percent of the ultimate strength of the metal for design purposes. See also Sec. 5-1.) Nomographs are provided by NFPA[31] to determine A_v for values of variables, including overshoot, that affect it; excerpts for dust-explosion venting in a 10-m^3 vessel are given in Table 5-2.

It is evident from Table 5-2 that there is wide variation in the required explosion venting area A_v depending on the values of the several variables that affect it. The following additional observations can be made:

1. The maximum vented explosion pressure p_v is not limited to the burst pressure of a rupture disk p_b, as is sometimes erroneously assumed.

2. Strong ignition sources require more explosion venting area than weak sources. Although strong ignition sources are uncommon, a strong source should be assumed if there is doubt about ignition strength.[31] There is uncertainty about turbulence. The degree of turbulence should be the same in the test and plant vessels for the cubic law to apply. Although the manner of dust-explosion testing inherently produces turbulence, there may be unknown turbulence; the bursting of the disk itself can create turbulence and enhanced explosion severity. Thus, the explosion-venting area should be as large as feasible.

TABLE 5-2 Venting of Dust Explosions inside a 10-m³ Vessel (Data from Ref. 31)

Ignition	Hazard class	p_v, kPa gage*	p_b, kPa gage	A_v, m^2
Weak	St-1	200	10	0.21
			50	0.24
		60	10	0.37
			50	0.83
	St-2	200	10	0.40
			50	0.44
		60	10	0.63
			50	1.5
Strong	St-1	200	10	0.40
			50	0.43
		60	10	0.70
			50	1.5
	St-2	200	10	0.60
			50	0.75
		60	10	1.0
			50	2.4

*Initial pressure = 101 kPa abs.

3. The larger the permissible p_v the smaller the required A_v. Thus, the largest practicable design pressure should be used to allow manageable explosion vent areas.

The Factory Mutual Engineering Corporation has published a table (Table 1) of unvented and vented explosion pressures for a myriad of dusts in their Loss Prevention Data Sheet 7-76, Combustible Dusts. Tests are performed in their 3.6-L dust-explosion test apparatus. (Vent ratios are $1/55$, $1/20$, and $1/11$ m^{-1}.) Factory Mutual accounts for the decrease in r_m and in required explosion venting area with increasing vessel volume as follows*:

a. For equipment with volumes up to 0.28 m^3, use the full pressures for venting ratios reported in Table 1.

b. For equipment with volumes from 0.28 m^3 to 2.8 m^3, design for either half the pressure reported in Table 1 using the full venting ratio, or use the full pressure and half the venting ratio.

c. For equipment with volumes over 2.8 m^3, design for one-quarter of the pressure reported in Table 1 using the full venting ratio, or use the full pressure and one-quarter of the venting ratio.

Bartknecht[11] indicated that long containers, such as silos, should have the entire top used for explosion relief (A_v/V = 1/height). Otherwise, the top may be damaged. Even so, the volume thus protected for a prescribed p_v may be limited. Strengthening the container may be required for safe use of a large volume; the dust cloud should be assumed to occupy the entire volume. These requirements may be impractical in many situations, and blanketing of the silo with inert gas usually is feasible if there is an explosion threat.

Gases of most solvents and saturated hydrocarbons have about the same S_u and therefore the same K_G as propane. Nomographs for weak ignition are provided by NFPA[31] for propane, natural gas, hydrogen, and coke gas. Examples of required A_v with propane for specified p_b and p_v are shown in Table 5-3. (K_G = 75 bar·m/s.)

As can be seen from the 10-m^3 entries in Table 5-3 and from Table 5-2, nonturbulent propane and comparable gases with weak ignition have about the same explosion-venting requirements as St-2 dusts with strong ignition. With saturated hydrocarbons and solvents, turbulence and strong ignition can increase explosion venting requirements still further. In that case the more stringent NFPA venting requirements for hydrogen should be used (K_G = 550 bar·m/s).[31]

High-pressure venting. As indicated by Eqs. (4-3) and (4-4), high initial pressure increases r_m compared with P_i = 101 kPa abs. Ferris[20] performed useful explosion-venting tests with methyl alcohol in a 1.9-m^3

*Quoted from Factory Mutual Engineering Corporation Data Sheet 7-76 by permission.

TABLE 5-3 Venting of Propane Explosions (Data from Ref. 31)

Vessel volume, m^3	p_v, kPa gage*	p_b, kPa gage	A_v, m^2†
5	200	10	0.30
		50	0.42
	60	10	0.71
		50	1.0
10	200	10	0.50
		50	0.73
	60	10	1.3
		50	1.9
50	200	10	1.4
		50	2.1
	60	10	3.8
		50	5.2
100	200	10	3.0
		50	4.0
	60	10	6.8
		50	9.0

*Initial pressure = 101 kPa abs.
†Stationary mixture when ignited; weak ignition.

spherical vessel. Initial conditions were with a temperature of 120°C and a pressure of 83 kPa gage. Fans were used to produce turbulence, and simultaneous ignition was provided at 16 locations in the vessel. Thus, the tests are representative of turbulent mixtures with strong ignition. (See the discussion of vessel geometry, volume, and shape in Sec. 4-3.) The ratio of vented explosion pressure to explosion vent A_v/V is shown in Fig. 5-5 for 15% methyl alcohol in air with a nominal burst pressure of the rupture disk of 138 kPa gage. (The C_{st} is 12.3% v/v, but 15% gives maximum explosion effects.) Extrapolated p_v for other volumes from Eq. (5-4) are shown by dashed lines in Fig. 5-5.

For corresponding conditions, Fig. 5-5 can be used for saturated hydrocarbons and solvents. Also, as developed by NFPA,[31] if the ratio of the vent-bursting absolute pressure to the initial gas absolute pressure is kept constant, and if vessel size and vent size are kept constant, the pressure developed during the venting of propane and similar gas combustion will vary as about the 1.5 power of the initial absolute pressure. For hydrogen the exponent ranges from 1.1 to 1.2. These exponents are applicable up to an initial pressure of approximately 400 kPa abs. An example, using a propane nomograph, is provided by NFPA.[31] Moreover, tests may be necessary to determine the required rupture-disk size. Experiments in a properly sized test vessel can be used to determine the size of a plant explosion vent with the aid of Eq. (5-4).

Runaway reactions in polymerization reactors may lead to simultaneous discharge of vapor and liquid. The emergency venting of these two-phase releases is a special problem that has been reviewed by several

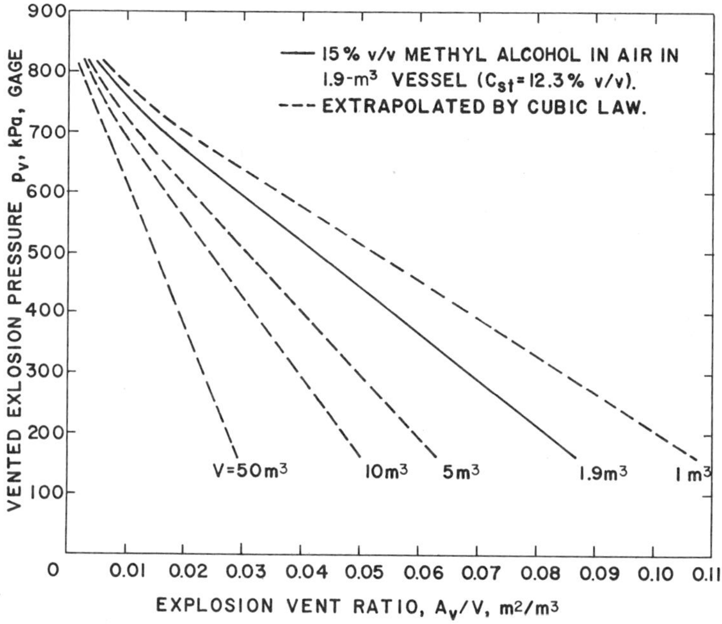

Fig. 5-5 Vented explosion pressure for 15% v/v methyl alcohol in air. Initial temperature and pressure of 120°C and 83 kPa gage. Initially, turbulent mixture with strong ignition. Nominal burst pressure of rupture disk = 138 kPa gage. *(Data of Ferris*[20] *for 1.9-m*^{3} *vessel.)* Dashed lines = extrapolated p_v for other volumes from Eq. (5-4).

investigators.[2,45–48] Factory Mutual Engineering Corporation Loss Prevention Data Sheet 7-49, 12-65, Emergency Venting of Vessels, provides further information. In addition, the Design Institute for Emergency Relief Systems of the American Institute of Chemical Engineers is implementing a research program on pressure relief involving two-phase flow.

In general, vents based on vapor flow alone from the heat released during a runaway polymerization are inadequate. Even with consideration of two-phase flow, a safety factor of 2 to 3 on area should be applied.[2,45] Also, the rupture-disk setting should be as close as possible to the operating pressure. Moreover, an inert-gas-purged knockout drum should be installed to catch the liquid if there are environmental or explosion hazards without collection.

Selection, Installation, and Maintenance of Rupture Disks

Selection. Relief valves ordinarily do not open fast enough to protect a vessel from the fast rate of pressure rise occurring in an explosion. A Bicera (British International Combustion Engine Research Association) crankcase explosion-relief valve, manufactured by the Penn-Troy Machine Company, Troy, Pennsylvania, is used to relieve crankcase explosions. As a desirable feature, it is equipped with an internal flame arrester to limit flame spread into the atmosphere. Enough valves must be installed to provide a minimum venting ratio A_v/V of 0.034 m^{-1} to limit explosion pressure to 170 to 205 kPa gage. Normally, though, a rupture disk is used to relieve explosions in vessels. A rupture disk is a pressure-relieving device consisting of a fragile disk, usually of metal, graphite, or plastic materials, held between special flanges and of such a thickness, diameter, shape, and material that it will rupture at a predetermined pressure. It does not reclose. Information on selection, installation, and maintenance of rupture disks is included in Refs. 31 and 49 to 52. The Factory Mutual Engineering Corporation has published a Loss Prevention Data sheet, Selection and Application of Rupture Disks, 12-46. Additionally, manufacturers of rupture disks, such as BS&B Safety Systems, Inc., Tulsa, Oklahoma, provide extensive information on their products in catalogs. The various types of rupture disks and some comments on them are listed in Table 5-4.

The ASME Boiler and Pressure Vessel Code, section VIII, division 1 states, "Every disk shall have a stamped bursting pressure at a specified disk temperature, guaranteed to burst within plus or minus 5 percent of its stamped bursting pressure." Elevated temperature has a large effect on the burst pressure of rupture disks; this effect varies with the disk material. For example, a type B (prebulged forward-acting solid metal) 316 stainless-steel disk of the BS&B Safety Systems, Inc., at 460°C bursts at about 75 percent of its rated pressure at 22°C. Manufacturers should be consulted on the maximum temperature and pressure and minimum pressure for application of a specified disk. Thus, it is important that the disk temperature be known; temperature measurements may be necessary.

Corrosion can cause premature failure of a disk. A wide selection of materials is supplied by disk manufacturers to combat corrosion. Also, linings are available for disks and safety heads to resist corrosion further. Possible corrosion on the downstream side of the disk should be considered; corrosion may occur from contaminants in the atmosphere or from process material left in the vent from a previous blow.[50]

Several accessories are available for use with rupture disks. Safety heads for holding the disks come in a wide range of types, sizes, materials, pressure, and temperature ratings. BS&B Safety Systems, Inc., supplies an insert type of safety-head assembly that rests between the studs of two American National Standards Institute (ANSI) flanges. This head can be removed and replaced without dismantling piping. They also supply jackscrews for separating flanges and eyebolts for handling on their safety-head preassemblies. A STA-SAF*

*STA-SAF is a trade mark of BS&B Safety Systems, Inc.

TABLE 5-4 Types of Rupture Disks

Type	Description	Comment
Prebulged forward-acting solid-metal disk	One-piece dished metal disk with pressure on the concave side (Fig. 5-6); vacuum support on concave side of disk may be needed if atmospheric pressure exceeds process pressure	Temperature and pressure have greatest effect on this disk; margin between operating and burst pressure must be greater than with other types
Composite rupture disk	Two or more separate components; top section is prebulged metal with radial or circumferential slots; bottom may be made from metal or plastic and is for sealing; pressure is on concave side (Fig. 5-7); vacuum support may be needed	Better suited to lower-pressure uses than solid metal disks and less affected by pressure variations
Reverse-buckling (RB) disk	Prebulged solid metal disk with pressure on convex side; knife blades on downstream side pierce disk when bulge reverses direction upon application of sufficient pressure; vacuum supports not required since disk metal is thicker than with prebulged disks	Closer margin between operating pressure and rated pressure of disk; no fragmentation; bolting torque critical with RB; consult manufacturer for use with normally liquid-filled systems; RB disks must be handled very carefully
Scored disk	Scored forward-acting disk is similar to prebulged metal disk but with score lines; it can achieve lower burst pressures than unscored disk; scored reverse-acting disk is similar to RB, but there are no knife blades; disk bursts along scored lines on reversal; no fragmentation (Fig. 5-8)	
Graphite disk	Flat piece of graphite is supported in special retainer; rated burst pressure is same in either direction; vacuum supports may be needed on disks rated below 101 kPa, gage	Graphite disks are used in low-pressure and pulsating-pressure services; resistant to most corrosive chemicals; fragile
Explosively operated disk	Disk ruptured by explosive, actuated manually or automatically by sensing	

TABLE 5-4 *(continued)*

	device that closes electric circuit to explosive; devices sense temperature, pressure, etc.	
Explosion panels	Membrane opens upon development of explosion pressure	Useful for low burst pressure; flexing may cause premature failure; if grid for support from pulsating pressure is used, it must be on process (not atmospheric) side of panel; otherwise, panel will burst at much higher than rated pressure

Fig. 5-6 Prebulged solid-metal rupture disk: *top,* unburst; *bottom,* burst. *(Type B, BS&B Safety Systems, Inc.)*

system incorporates a scored reverse-buckling rupture disk in a mated safety head, as shown in Fig. 5-9.

This system assures correct installation and allows field inspection and cleaning. The pins on the inlet assembly flange ensure that the disk is installed correctly in the safety head. The J bolt on the inlet

Fig. 5-7 Composite rupture disk: *top,* unburst; *bottom,* burst. *(Type D, BS&B Safety Systems, Inc.)*

flange mates with a drilled hole in the inlet companion ANSI flange to ensure installation in the correct flow direction. Thus, the J bolt and pins help eliminate incorrect installation. Even if the disk is installed incorrectly with the concave side to process pressure, it will fail as a scored, forward-acting disk but at a higher pressure. With stainless steel, for example, this higher pressure is a maximum of twice the rated burst pressure.

An automated disk changer is supplied by FMC Corporation, Fluid Control Division, Houston, Texas.[53] The assembly consists of a sliding-gate valve with a rupture disk mounted in each of two openings. A piston actuator moves the burst disk out of line, and the new disk slides into position. The ruptured disk can be replaced later, and thus downtime is reduced. The changer accommodates carbon or metal disks and can be made from carbon steel or corrosion-resistant materials.[53] It is available in sizes from 50 to 305 mm diameter.

Installation and maintenance. Walls of buildings usually cannot withstand much pressure. *Do not vent explosions into closed rooms;* unburned, vented material can explode in the room, and secondary dust explosions may occur if settled dust is dislodged from the primary explosion. More-

Fig. 5-8 Scored reverse-buckling rupture disk: *top,* unburst; *bottom,* burst. *(Type S-90, BS&B Safety Systems, Inc.)*

over, the end of the vent must be so located and directed that it does not pose a threat to personnel; the flame may extend 25 m or more from the vent.

A rupture disk is a fail-safe pressure-relief device, but it must be installed and maintained properly to preserve this feature. Fires may follow an explosion; in some cases installation of fire-protection equipment should be considered. Also, negative and perhaps damaging pressure can occur in equipment after an explosion by cooling if an explosion panel, such as a swinging door, recloses. Stops, or similar devices, may be used on the vent closures to prevent vacuum.

If problems occur with rupture disks or safety heads, they are most often due to incorrect installation.

Installation checklist. Implementation of items in the following checklist is needed to provide dependable performance of rupture disks and safety heads.

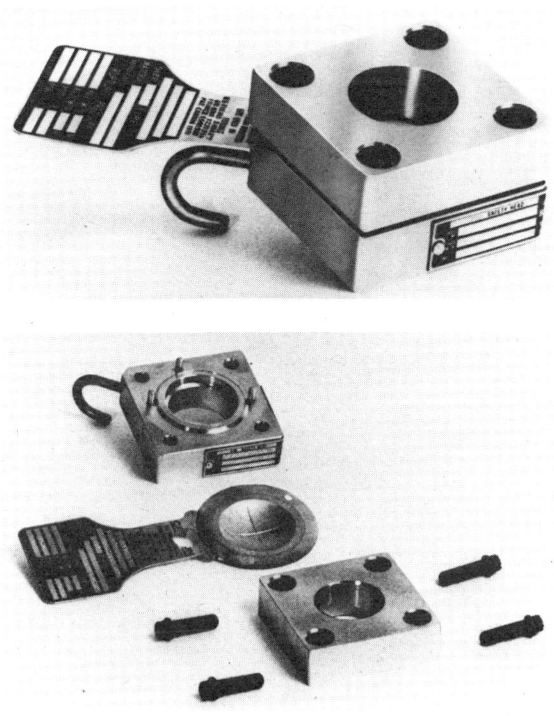

Fig. 5-9 STA-SAF system; 50-mm SRB-7RS safety head with S-90 scored reverse-buckling rupture disk: *top,* assembled; *bottom,* unassembled. *(BS&B Safety Systems, Inc.)*

- ✓ Install rupture disks in accordance with directions of the manufacturer.
- ✓ Handle disks carefully in storage and installation. Do not use damaged disks.
- ✓ Use disk of proper pressure rating.
- ✓ Install in right direction.
- ✓ Install vacuum support on the right side of disk if the support is not attached by the manufacturer.
- ✓ Follow torque recommendations of the manufacturer. Misalignment and improper torquing may result in premature failure of the disk.
- ✓ Keep flange surfaces clean.
- ✓ Confirm that the proper disk has been correctly installed on the right equipment.
- ✓ Replace disks periodically, as best determined by plant experience.

Also, poor corrosion experience may require a different disk or lining material.

- ✓ Check to see if plugging is possible, e.g., from freezing or polymerization. In addition, filter bags in dust-control equipment may interfere with venting. A way to counter this effect is to make provision for ample open space at the vent.
- ✓ Restrain doors and the like that are used as explosion vents with chains to prevent them from becoming missiles.

Duct tips. A discharge duct is needed to expel explosion products to the outdoors. The effect of duct length on vented explosion pressure has been studied by Donat,[19] Hartmann and Nagy,[23] Palmer,[32] and Wiekema et al.,[54] among others. A duct imposes pressure drop and increases vented explosion pressure compared with unrestricted venting. "In practice, this means that containers equipped with bursting disks and vent ducts must possess greater strength than containers with bursting disks of the same size but not equipped with vent ducts."[19] Tests have been performed, for example, in a 1-m^3 vessel with p_b from 20 to 50 kPa and with a duct diameter of 0.35 m; for a duct length-to-diameter ratio of more than about 20, pressures can be approximately 3 times higher in the vessel with the duct ($p_v \approx 300$ kPa) than without the duct ($p_v \approx 100$ kPa).[54] More research is needed, however, to assess the effect of duct length more quantitatively. Bends, too, may impose pressure drop. Thus, the ducting should be short and free from bends. Some additional reminders for installation of ducts for relief of explosions in vessels follow:

- ✓ Install rupture disk on vessel, not on end of duct.
- ✓ Make the area of all duct and piping from the vessel at least as large as the area of the rupture disk. There must be no restrictions in the line.
- ✓ Construct the duct to withstand a pressure at least as high as the maximum vented explosion pressure in the vessel.
- ✓ Brace for reaction forces. A baffle plate, such as supplied by BS&B Safety Systems, Inc., may be installed to absorb recoil when the safety head is vented directly to atmosphere. (A *baffle plate* is a flat plate directly connected to the safety head. Enough room must be allowed between the baffle plate and safety head to avoid impeding the flow.)
- ✓ Monitor and/or vent the pressure between two disks in series if a double-disk assembly is installed. Otherwise, a leak in the upstream disk could cause that disk to burst at a higher than rated pressure. [Compression of air between two disks if the upstream disk fails prematurely might cause high enough temperature to ignite the released gas. This possible effect has to be considered if a double-

disk assembly is installed to save product and/or to decrease downtime. As an example, assume that the upstream disk fails prematurely at an operating absolute pressure of 1000 kPa. Adiabatic compression of air ($\kappa = 1.4$) at 101 kPa abs in the space between the two disks will raise the air temperature from 25 to 301°C according to Eq. (3-6).]

✓ Provide for proper collection and disposal if environmental or flammability problems can occur from the discharged materials (see Fig. 6-4).

✓ Rupture disk capacities determined in the laboratory under conditions of clean failure and zero duct length may be reduced by as much as 50% in plant applications, if the combined resistance of the failed rupture disk and the duct is 150–200 duct diameters.

References

1. Chuss, R., *Pressure Vessels,* 5th ed., McGraw-Hill, New York, 1977.

2. Duxbury, H. A., "Gas Vent Sizing Methods," *Chem. Eng. Progr. 10th Loss Prev. Symp., Kansas City, 1976,* pp. 147–150.

3. Charney, M., and F. K. Lawler, "Stops Explosions after They Start," *Food Eng.,* vol. 39, no. 10, pp. 82–85, October 1967.

4. Hammond, C. B., "Explosion Suppression: New Safety Tool," *Chem. Eng.,* vol. 68, no. 26, pp. 85–88, Dec. 25, 1961.

5. Martin, A. J., "Keep the Pressure in the Can," *Aerosol Age,* vol. 24, no. 2, pp. 26, 28, 30–32, 56, 57, February 1979.

6. National Fire Protective Association, Standard on Explosion Prevention Systems, *NFPA* 69, Boston.

7. Bartknecht, W., "Explosion wird in Millisekunden unterdruckt," *Chem. Ind.,* vol. 29, no. 7, pp. 393–395, July 1977.

8. Anthony, E. J., "The Use of Venting Formulae in the Design and Protection of Building and Industrial Plant from Damage by Gas or Vapour Explosions," *J. Hazard. Mater.,* vol. 2, no. 1, pp. 23–49, December 1977.

9. Bartknecht, W., "Report on Investigations on the Problem of Pressure Relief of Explosions of Combustible Dusts in Vessels," *Staub Reinhalt. Luft,* vol. 34, no. 11, pp. 289–300, November 1974.

10. Bartknecht, W., "Report on Investigations on the Problem of Pressure Relief of Explosions of Combustible Dusts in Vessels," *Staub Reinhalt. Luft,* vol. 34, no. 12, pp. 358–361, December 1974.

11. Bartknecht, W., "Explosion Pressure Relief," *Chem. Eng. Prog. 11th Loss Prev. Symp., Houston, 1977,* pp. 93–105.

12. Bradley, D., and A. Mitcheson, "The Venting of Gaseous Explosions in Spherical Vessels, I: Theory," *Combust. Flame,* vol. 32, pp. 221–236, July 1978.

13. Bradley, D., and A. Mitcheson, "The Venting of Gaseous Explosions in

Spherical Vessels, II: Theory and Experiment," *Combust. Flame,* vol. 32, pp. 237–255, July 1978.

14. Burgoyne, J. H., and M. J. G. Wilson, "The Relief of Pentane Vapour–Air Explosions in Vessels," *Inst. Chem. Eng. Symp. Ser.* 7, *Proc. Symp. Chem. Process Hazards Spec. Ref. Plant Des., 1960,* pp. 25–29.

15. Cousins, E. W., and P. E. Cotton, "Design Closed Vessels to Withstand Internal Explosions," *Chem. Eng.,* vol. 58, no. 8, pp. 133–137, August 1951.

16. Creech, M. D., "Combustion Explosions in Pressure Vessel Protected with Rupture Disks," *Trans. ASME,* vol. 63, no. 7, pp. 583–588, October 1941.

17. Decker, D. A., "Explosion Venting Guide," *Fire Technol.,* vol. 7, no. 3, pp. 219–223, August 1971.

18. Donat, C., "Selection and Dimensioning of Pressure Relief Devices for Dust Explosions," *Staub Reinhalt. Luft,* vol. 31, no. 4, pp. 17–25, April 1971.

19. Donat, C., "Pressure Relief as Used in Explosion Protection," *Chem. Eng. Prog. 11th Loss Prev. Symp., Houston, 1977,* pp. 87–92.

20. Ferris, T. V., "The Explosion of Methanol-Air Mixtures at Above Atmospheric Conditions," *Chem. Eng. Prog. 8th Loss Prev. Symp., Philadelphia, 1974,* pp. 15–19.

21. Gibson, N., and G. F. P. Harris, "The Calculation of Dust Explosion Vents," *Chem. Eng. Prog.,* vol. 72, no. 11, pp. 62–67, November 1976.

22. Harris, G. F. P., and P. G. Briscoe, "The Venting of Pentane Vapour–Air Explosions in a Large Vessel," *Combust. Flame,* vol. 11, pp. 329–338, August 1967.

23. Hartmann, I., and J. Nagy, "Venting Dust Explosions," *Ind. Eng. Chem.,* vol. 49, no. 10, pp. 1734–1740, October 1957.

24. Heinrich, H. J., "Bemessung von Druckentlastungsöffnungen zum Schutz explosiongefährdeter Anlagen in der chemischen Industrie," *Chem. Eng. Tech.,* vol. 38, no. 11, pp. 1125–1133, November 1966.

25. Heinrich, H. J., and R. Kowall, "Results of Recent Pressure Relief Experiments in Connection with Dust Explosions," *Staub Reinhalt. Luft,* vol. 31, no. 4, pp. 10–17, April 1971.

26. Heinrich, H. J., and R. Kowall, "On the Course of Pressure-Relieved Dust Explosions with Ignition through Turbulent Flames," *Staub Reinhalt. Luft,* vol. 32, no. 7, pp. 22–27, July 1972.

27. Maisey, H. R., "Gaseous and Dust Explosion Venting, Part 1," *Chem. Process Eng. (Lond.),* vol. 46, no. 10, pp. 527–535, and 563, October 1965.

28. Maisey, H. R., "Gaseous and Dust Explosion Venting, Part 2," *Chem. Process Eng. (Lond.),* vol. 46, no. 12, pp. 662–672, December 1965.

29. Morton, V. M., and M. A. Nettleton, "Pressures and Their Venting in Spherically Expanding Flames," *Combust. Flame,* vol. 30, no. 2, pp. 111–116, 1977.

30. Munday, G., "The Calculation of Venting Areas for Pressure Relief of Explosions in Vessels," *Inst. Chem. Eng. Symp. Ser.* 15, *Proc. 2d Symp. Chem. Process Hazards Spec. Ref. Plant Des., 1963,* pp. 46–54.

31. National Fire Protection Association., Guide for Explosion Venting, *NFPA* 68, Boston, 1978.

32. Palmer, K. N., "The Relief Venting of Dust Explosions in Process Plant," *Inst. Chem. Eng. Symp. Ser.,* 34, *Major Loss Prev. Process Ind., 1971,* pp. 142–147.

33. Palmer, K. N., *Dust Explosions and Fires,* Chapman & Hall, London, 1973.

34. Palmer, K. N., "Relief Venting of Dust Explosions," *Chem. Eng. Prog.,* vol. 70, no. 4, pp. 57–61, April 1974.

35. Rasbash, D. J., and Z. W. Rogowski, "Relief of Explosions in Duct Systems," *Inst. Chem. Eng. Symp. Ser.* 7, *Proc. Symp. Chem. Process Hazards Spec. Ref. Plant Des., 1960,* pp. 58–68.

36. Rasbash, D. J., and Z. W. Rogowski, "Gaseous Explosions in Vented Ducts," *Combust. Flame,* vol. 4, pp. 301–312, December 1960.

37. Rogowski, Z. W., and D. J. R. Rasbash, "Relief of Explosions in Propane-Air Mixtures Moving in a Straight Unobstructed Duct," *Inst. Chem. Eng. Symp. Ser.* 15, *Proc. 2d Symp. Chem. Process Hazards Spec. Ref. Plant Des., 1963,* pp. 21–28.

38. Runes, E., "Explosion Venting," *Chem. Eng. Prog. 6th Loss Prev. Symp., San Francisco, 1972,* pp. 63–67.

39. Schwab, R. F., and D. F. Othmer, "Dust Explosions," *Chem. Process Eng. (Lond.),* vol. 45, no. 4, pp. 165–174, April 1964.

40. Simonds, W. A., and P. A. Cubbage, "The Design of Explosion Reliefs for Industrial Drying Ovens," *Inst. Chem. Eng. Symp. Ser.* 7, *Proc. Symp. Chem. Process Hazards Spec. Ref. Plant Des., 1960,* pp. 69–77.

41. Yao, C., "Explosion Venting of Low-Strength Equipment and Structures," *Chem. Eng. Prog. 8th Loss Prev. Symp., Philadelphia, 1974,* pp. 1–9.

42. Grummer, J., M. E. Harris, and V. R. Rowe, "Fundamental Flashback, Blowoff, and Yellow-Tip Limits of Fuel Gas-Air Mixtures," *U.S. Bur. Mines Rep. Invest.* 5225, 1956.

43. Crane Co., Flow of Fluids through Valves, Fittings and Pipe, *Techn. Pap.* 410, Chicago, 1978.

44. Nagy, J., E. C. Seiler, J. W. Conn, and H. C. Verakis, "Explosion Development in Closed Vessels," *U.S. Bur. Mines Rep. Invest.* 7507, April 1971.

45. Boyle, W. J., Jr., "Sizing Relief Area for Polymerization Reactors," *Chem. Eng. Prog.,* vol. 63, no. 8, pp. 61–66, August 1967.

46. Harmon, G. W., and H. A. Martin, "Sizing Rupture Discs for Vessels Containing Monomers," *Chem. Eng. Prog. 4th Loss Prev. Symp., Atlanta, 1970,* pp. 95–103.

47. Huff, J. E., "Computer Simulation of Polymerizer Pressure Relief," *Chem. Eng. Prog. 7th Loss Prev. Symp., New York, 1973,* pp. 45–57.

48. Kneale, M., and J. S. Binns, "Relief of Runaway Polymerizations," *Inst. Chem. Eng. Symp. Ser.* 49, *Proc. 6th Symp. Chem. Process Hazards Spec. Ref. Plant Des. 1977,* pp. 47–52.

49. Block, B., "Emergency Venting for Tanks and Reactors," *Chem. Eng.,* vol. 69, no. 2, pp. 111–118, Jan. 22, 1962.

50. Kayser, D. S., "Rupture Disc Selection," *Chem. Eng. Prog. 6th Loss Prev. Symp., San Francisco, 1972,* pp. 82–87.

51. Nagy, J., J. E. Zeilinger, and I. Hartmann, "Pressure-Relieving Capacities of

Diaphragms and Other Devices for Venting Dust Explosions," *U.S. Bur. Mines Rep. Invest.* 4636, January 1950.

52. "Symposium on Bursting Discs," *Trans. Inst. Chem. Eng.*, vol. 31, pp. 113–167, 1953.

53. Anon., "Reactor Shutdown Eliminated by Push-Button Changer for Ruptured Pressure Disc," *Chem. Process (Chicago)*, vol. 39, no. 1, p. 79, January 1976.

54. Wiekema, B. J., H. J. Pasman, and T. M. Groothuizen, "The Effect of Tubes Connected with Pressure Relief Vents," pp. IV-223–IV-231, in *Proc. 2d Int. Symp. Loss Prev. Saf. Promot. Process Ind.* DECHEMA, Deutsche Gesellschaft für chemisches Apparatewesen, Frankfurt, 1978.

Additional Reference

Minors, C. (ed.), "The Safe Venting of Chemical Reactors," *Inst. Chem. Eng., NW Branch, Symposium Papers, 1979,* No. 2.

6

Atmospheric Releases

Off-gases often have to be released to atmosphere. Since they may cause toxicity and/or the threat of explosion, these emissions require adequate controls to prevent such problems. *Close scrutiny of possible atmospheric concentrations relative to potentially toxic and flammable concentrations is essential.* Also, governmental environmental and occupational health regulations must be implemented. Several types of problems can develop from the continuous or emergency emissions of flammable gases. These problems and the factors requiring consideration in their solution are covered in this chapter.

6-1 Releases Containing Air

Flashback

Flashback into a pipe from which a flammable mixture issues can occur if the stream velocity near the wall falls below the burning velocity S_u. The diameter of the tube must be larger than the quenching distance for flashback to occur. Ordinarily, quenching distances are only a few millimeters, i.e., much smaller than industrial vent diameters. (The *quenching distance* is the minimum spacing of walls of a channel through which a given flame can propagate in a quiescent mixture.[1]) In turbulent flow, the conditions near the center of stream may lead to flashback there as well.[2] Industrial vent emissions are preponderantly turbulent, with Reynolds numbers greater than about 2000. Most experimental work on flashback has been done with laminar flow. Turbulence increases the burning velocity, but far too little research has been done on turbulent flashback. Flashback speeds are highest at about the stoichiometric concentration in air. Flammable discharges can be ignited accidentally by an external ignition source. Thus, if flammable-gas–air mixtures must be emitted, their discharge speed should exceed turbulent flashback speeds whenever possible to help prevent flame propagation into plant equipment. In this case, the flame will burn only in the open air at the vent tip. Generally, an average vent gas velocity of 4.5 m/s will suffice for saturated hydrocarbon-air and solvent-air mixtures in vents up to about 300 mm diameter. (Hydrogen-air mixtures can flash back below 13 m/s with vent diameters up to 38 mm.[3])

Flame Arresters

Whatever the vent-gas exit velocity, a flame arrester should be installed if a flammable gas-air mixture is emitted to atmosphere. The main function of an arrester is the absorption of heat, thereby preventing passage of flame. Quenching distances are the maximum separation of solid surfaces at which heat extraction by those surfaces prevents flame propagation.

> Arresters usually consist either of an aggregation of parallel small channels or a maze of small channels and are intended to offer minimal resistance to gas flow. Common types of arresters are fabricated from wire gauzes, perforated sheet, crimped metal ribbon, and sintered metals. Arresters may also consist of towers packed with pebbles, beads, or Raschig rings. Typically, arresters are installed in such locations as solvent recovery systems, vent pipes of storage tanks for flammable liquids, and feedlines for premixed gases fed to burners and furnaces.[4]

The first use of a flame arrester was the wire gauze in Davy's miner's safety lamp, developed early in the nineteenth century. Hydraulic arresters are also used, but it must be assured that the liquid, normally water, breaks up the gas stream into discrete bubbles to produce discontinuity of the gas stream; cooling the flame is not the primary function of hydraulic arresters. A flame arrester is shown in Fig. 6-1.

Storage tanks for flammable liquids are usually equipped with a flame arrester with no lead-off duct. Often an arrester is used with a conservation vent in a single device in these installations. In-line arresters are used in vent pipes. The length of pipe can affect the arrester's ability to stop flame because flame speeds increase with run-up distance. Underwriters' Laboratories, Inc., list maximum permissible vent-pipe lengths from the arrester to the open pipe end for several commercial flame arresters.

Flame arresters provide a means of reducing potential explosion hazards, yet problems can lurk that need to be considered in design, operation, and maintenance.

Pressure drop. Flame arresters impose a pressure drop that may not be tolerable. Pressure drop can be reduced by increasing the diameter of the arrester or by installing two or more arresters in parallel.

Clogging. Interstices of arresters may become blocked for a variety of reasons. They should not be used with dusty emissions. Condensation,

Fig. 6-1 Flame arrester, 75-mm pipe size, schedule 40. Exploded view on bottom. Gas flow is from left to right. *(Model F-1, The C. M. Kemp Manufacturing Company.)*

freezing, crystallization, polymerization, and corrosion can cause clogging. Steam-jacketed flame arresters are available commercially. Corrosion-resistant materials of construction may be necessary.

Overheating. If a flammable gas-air mixture flows during and after ignition, flame may stabilize on the arrester and flash back through the arrester. Flame arresters are not intended to resist the flame of continuous-burning mixtures for long periods of time. A thermocouple and alarm can be installed to warn of impending danger. Also, commercially available fire checks are designed to shut off the flow of combustible gas mixtures by thermally operated shutoff valves in the event of flashback. An alternate vent may have to be provided in these circumstances.

Inspection and maintenance. Regular inspection and maintenance is required. Flame arresters should be installed in locations to permit easy inspection and maintenance.

6-2 Releases without Air

As often as not, combustible-gas releases to atmosphere do not contain air. Large quantities of flammable gases may be emitted through pressure relief valves to flares or high stacks. Air exclusion by an adequate inert-gas purge, prevention of flammable concentrations at potential ignition sources, and an overall safe disposal system are imperative.

Inert-Gas Purges

A flame arrester is not essential if the emission does not contain air. A continuous inert-gas purge to exclude potentially hazardous intrusion of ambient air down the vent is often advisable, however. Alternatively, flammable gases, e.g., natural gas, may be used. Purge rates to keep oxygen at "safe" values 7.5 m and more down a vent tip are shown in Fig. 6-2. (These rates are not enough to prevent flashback if sufficient air enters upstream of the vent and ignition occurs at the vent tip.)

Safe oxygen in Fig. 6-2 depends on molecular weight of purge gas, as follows:

MW of purge gas	"Safe" oxygen, % v/v
2	3
4	3
6	4
8	5
>8	6

These oxygen concentrations are less than the MOCs in Table 2-4 except for carbon monoxide. Purges should not be less than about 0.3 m^3/h, whatever the diameter. Purge rates should be measured, and an alarm should be provided to signal low flow of purge gas. Alternatively, it may be advisable to monitor for oxygen in special cases.

Hot emissions will cool rapidly in an idle stack when flow ceases, particularly in a rain. If the stack is not insulated and traced, additional purge gas will be necessary to compensate for the volume contraction by cooling. In such cases, it is prudent to sweep out the idle flammable gas as quickly as practicable. The purge can then be reduced to normal rate.

Heavy gases, such as CO_2, provide superior inerting of vent stacks in regard to air entry through the vent tip; a stack-gas speed of less than an estimated 0.3 mm/h CO_2 was required to keep air out of a 590-mm-diameter stack (< 80 mL/h).[5] Steam has also been used for purging stacks, but condensation can result in ineffective inerting. A visible steam plume does not assure adequate inerting. In fact, the appearance of a plume may be caused by condensation in the stack. Temperatures of saturated steam-air mixtures at heights of concern should exceed the

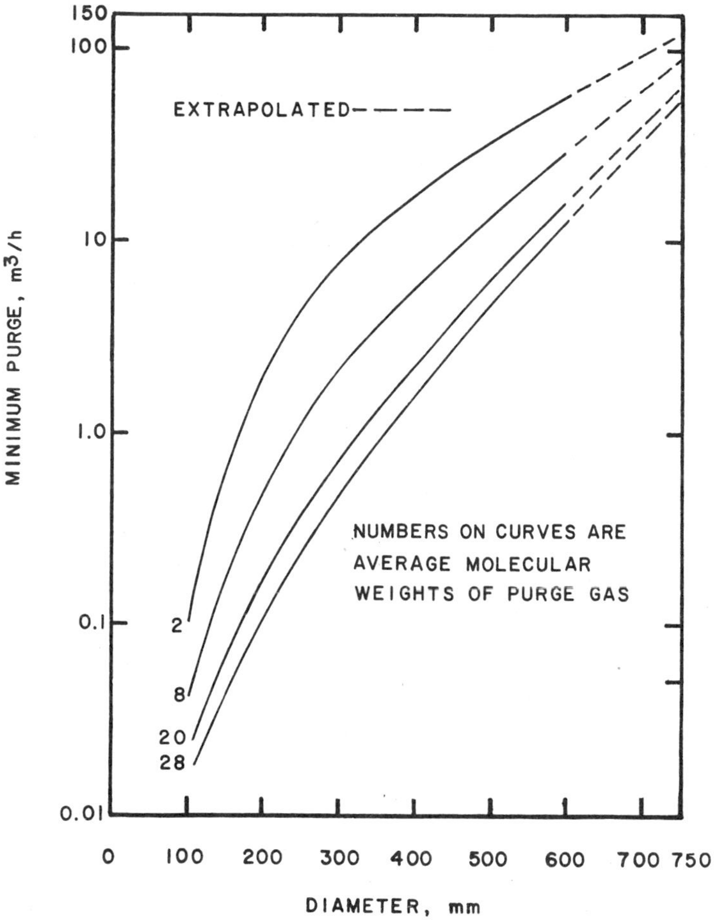

Fig. 6-2 Minimum adequate purge rates for stacks. *(After Husa,[5] by permission.)*

temperature values in Table 2-6 for appropriate O_2. Husa[5] has indicated that 0.67 kg/h per millimeter of diameter is the minimum adequate steam purge for stacks about 45 m high in any weather. ($v_s D = 400$ at 100°C.)

Behavior of Dense Stack Gases

Dense stack gases can descend rapidly, and this sinking can cause flammable concentrations at potential ignition sources. (*Vapor density* is often used to denote the specific gravity of a vapor relative to air. Thus, the vapor density of toluene is 3.1. At 25°C the vapor pressure of toluene is 3.80 kPa and, accordingly, the specific gravity cannot exceed 1.08. *Saturated vapor density* and *vapor-air density* are better terms.)

Bodurtha[6] was the first to study the behavior of dense stack gases. Photographs of plume behavior are shown in Fig. 6-3.

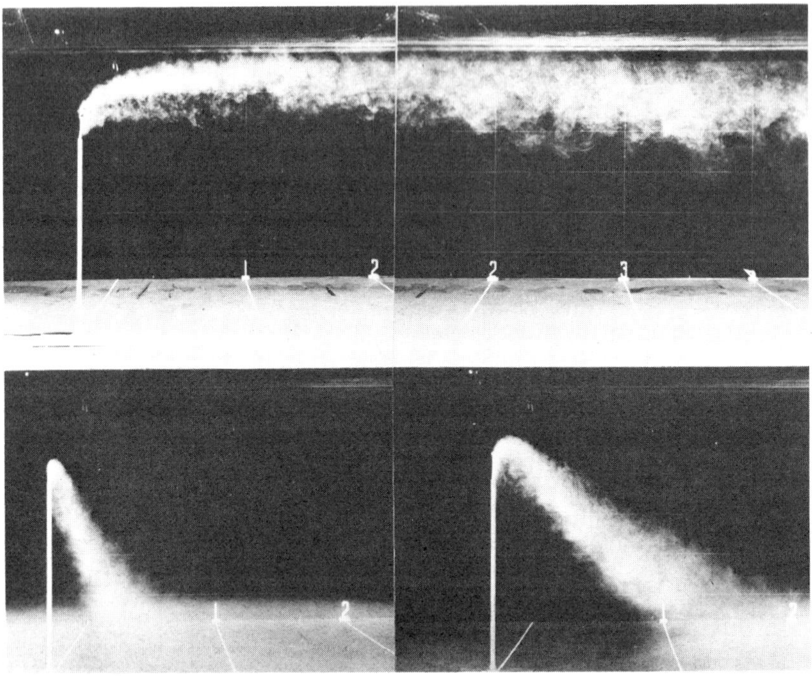

Fig. 6-3 Plume behavior in wind-tunnel tests. Full-scale conditions: $u = 1.34$ m/s, $D = 610$ mm, $v_s = 6.10$ m/s, $h_s = 30.5$ m: *upper,* SG = 1 (air); *lower left,* SG = 5.17; *lower right,* SG = 1.52. *(From Bodurtha,[6] by permission.)*

Hoot and Meroney[7] also studied the behavior of negatively buoyant stack gases in a wind tunnel. For "worst-case" conditions in light winds, wherever the plume touches the ground they determined that

$$\chi_m = 3.44\chi_s\left(\frac{D}{2H + h_s}\right)^{1.95} \frac{v_s}{u} \times 10^{-6} \tag{6-1}$$

where

$$H = 1.32D \left(\frac{v_s}{u}\right)^{1/3} (\mathrm{SG}^{1/3})\mathrm{Fr}^{2/3} \times 10^{-3} \tag{6-2}$$

and

$$\mathrm{Fr} = \frac{31.62 v_s}{\sqrt{gD\ (\mathrm{SG} - 1)/\mathrm{SG}}} \tag{6-3}$$

SG is for the total mixture, not just the heavy component. These equations should be used only if v_s at the vent tip is below sonic speed. The sampling time for concentration measurements was a few minutes. Momentary concentrations will be higher by a factor of about 5. Therefore, χ_m should be divided by about 5 when determining the stack height to prevent flammable concentrations at h_s m below the vent tip ($\chi_m = L/5$). Concentration maxima will occur with low wind speeds, and it is suggested that 1 m/s be used as the minimum wind speed. Also, highest concentrations will occur with SG = 2, that is, A = 1.59. Furthermore, Eqs. (6-1) and (6-2) can be maximized with respect to v_s, as shown in Appendix C. The resulting critical stack-gas exit velocity $v_{s,\,\text{crit}}$ is

$$v_{s,\,\text{crit}} = \frac{3.73D^{0.684}(5\chi_s/L)^{1.05} \times 10^{-3}}{u^{0.368}A^{2.05}} \tag{6-4}$$

Accordingly, at $v_{s,\,\text{crit}}$

$$h_s = \frac{43.8D^{1.35}(5\chi_s/L)^{1.05} \times 10^{-6}}{u^{0.702}A^{1.05}} \tag{6-5}$$

Compatible with pressure-drop considerations, vent diameters should be minimized to permit safe use of lower vent heights. [Using a minimum diameter has the further advantage of providing a high discharge speed that can extinguish (blow off) a flame if the emission is ignited accidentally.] The American Petroleum Institute indicates that vents are generally sized for an exit velocity of at least 152 m/s at the maximum relief rate.[8] To reduce pressure drop, a line size greater than the tip diameter can be used. The tip, then, can be nozzled to provide a greater exit velocity. Potential noise problems, however, should be considered in this selection. *Vents must be directed straight up to reap the benefits of the rise of the plume.* Turned-down vent tips cause potentially hazardous concentrations. (A weep hole at the low point of the discharge line can be used to drain condensate and rainwater.) In addition, ignition of an emission from a low, turned-down vent could cause flame impingement on process equipment or a tank. BLEVEs (boiling-liquid-expanding-vapor explosions) occur when a pressure vessel is heated so that the metal loses its strength and bursts[9]; they have caused considerable damage. Greatest potential for weakening of the metal and ultimate rupture of the vessel occurs with flame contact above the liquid level. With fire below the liquid level, the heat of vaporization provides a heat sink, as with a teakettle, thereby preserving the integrity of the metal and tank.

A method of handling a propane emission from a pressure relief valve is given in the following example. (Maximum emission rate = 40,000 kg/h.)

Example

$$\begin{aligned} D &= 203 \text{ mm} \\ T_s &= 298 \text{ K} \\ u &= 1 \text{ m/s} \\ v_s &= 192.5 \text{ m/s} \\ \text{SG} &= 1.52 \\ A &= 1.64 \\ L &= 2.2\% \text{ v/v} \\ \chi_s &= 100\% \text{ v/v} \end{aligned}$$

From Eq. (6-4)

$$v_{s,\,\mathrm{crit}} = \frac{3.73(203)^{0.684}(500/2.2)^{1.05}}{(1)^{0.368}(1.64)^{2.05}} \times 10^{-3} = 15.3\ \mathrm{m/s}$$

The actual stack-gas speed $v_{s,\,\mathrm{crit}}$ at the maximum emission rate of 40,000 kg/h is considerably more than the critical speed of 15.3 m/s. Peak concentrations will occur at the critical stack-gas exit speed. Efflux speeds will pass through $v_{s,\,\mathrm{crit}}$, and it is generally judicious to base required stack height on $v_{s,\,\mathrm{crit}}$. Thus, from Eq. (6-5)

$$h_s = \frac{43.8(203)^{1.35}(500/2.2)^{1.05}}{(1)^{0.702}(1.64)^{1.05}} \times 10^{-6} = 10.1\ \mathrm{m}$$

If the actual stack-gas exit velocity does not reach calculated $v_{s,\,\mathrm{crit}}$, Eqs. (6-1) and (6-2) should be used to determine h_s. The stack height required for tolerable radiation in the event of accidental ignition is often more than that required to prevent flammable concentrations.[8]

Pressure Relief Valves

Safety valves and safety relief valves are used to discharge gases and vapors to prevent potentially damaging overpressure to process equipment. Proper design of the relief system and provision for safe disposal of the released material is vital for safe operation. A guide for pressure relief systems and a standard for pressure relief valves are covered in Refs. 8 and 10, respectively. Key points that require emphasis for design of a relief valve and connecting piping are *adequate provisions for thermal expansion and for reaction forces resulting from discharge.* Required height and other geometric items for the discharge line have been covered in the previous section. Means of minimizing ignition are discussed in Chap. 3.

It is not generally possible to rule out ignition. Thus, questions inevitably arise about the potential for damaging pressure if accidental ignition does occur. With adequate vent heights, damaging overpressure from accidental ignition of relief-valve vapor emissions has not been experienced. The reasons for this are at least threefold:

1. Lack of strong igniters near the vent tip.
2. Rapid dilution of the effluent with air. This results in only a small amount of gas in the flammable range at any instant, even though the emission rate may be high.[11]
3. Rapid decrease in overpressure with distance from the vent tip, as shown in Fig. 4-10.

There has been only one reported incident of destructive pressure in the open air from ignition of a vent discharge.[11] Hydrogen was being emitted from a low (4.9-m) vent; a higher vent would have decreased the superficial physical damage that did occur. Ignition of hydrogen in a mishap of a lighter-than-air craft over Hull, England, in 1921 shattered thousands of windows in a 3-km radius. That release was unlike a vent release, however, because of the instantaneous enormous volume of hydrogen. Hydrogen emissions from relief valves may require more attention relative to overpressure from accidental ignition than other gases; glass breakage is the most common problem.

Sometimes liquid is released into the discharge line from a pressure relief valve. In such cases it is essential to install a knockout drum to separate the liquid. Otherwise, ignition of large liquid particles could cause burning drops to fall on the plant (Roman-candle effect). (Also, in some cases condensation in the vent stack can form liquid drops if the vent is not insulated and traced, even with a knockout drum.) Air should be purged from the knockout drum to prevent occurrence of flammable concentrations within it, should, say, a relief-valve leak. Once the air is displaced, a continuous inert-gas purge determined from Fig. 6-2 can be used.

A schematic diagram of a disposal system incorporating the features previously described is shown in Fig. 6-4.

Frequently, releases from a relief valve are directed to a flare. Several of the safety factors described earlier in this chapter should be considered in the design of a flare stack. Details of flare-stack design are adequately covered in Ref. 8. Ground flares may pose increased and sometimes unacceptable safety risks due to the relatively high concentrations of flammable materials that can occur in event of flame failure.

Continuous pilots on a flare stack provide a constant ignition source. It therefore is imperative to keep air out of a flare system. (On rare occasions flammable gas-air mixtures have been flared. Special procedures and equipment are provided in these cases.) Commercial seals of

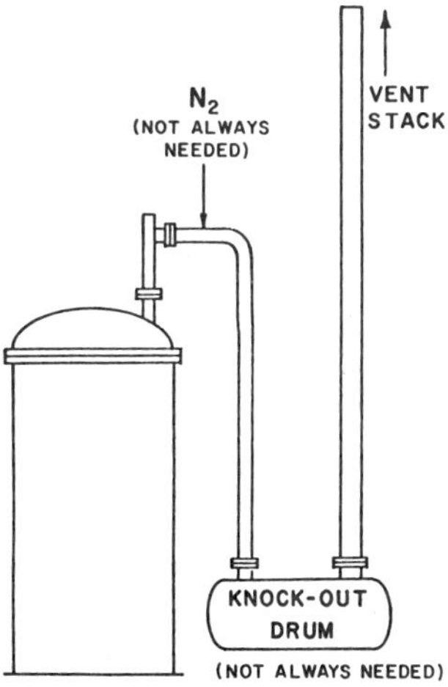

SAFETY RELIEF VALVE CHECKLIST

- MINIMIZE EXIT DIAMETER OF VENT STACK
- DIRECT VENT TIP STRAIGHT UP
- DO NOT DISCHARGE INTO BUILDING
- CONSIDER
  - o REACTION FORCES. PROVIDE SUPPORT FOR PIPING AND VALVES.
  - o NOISE
  - o THERMAL EXPANSION
  - o THERMAL SHOCK OF COLD LIQUIDS
  - o LIGHTNING PROTECTION
- SAFE HEIGHT OF VENT STACK BASED ON
  - o FLAMMABLE CONCENTRATIONS
  - o AIR POLLUTION
  - o THERMAL RADIATION

Fig. 6-4 Schematic diagram and checklist of a disposal system from a safety relief valve.

the John Zink Company and the National Airoil Burner Co., Inc., are available for installation in a stack to reduce air entry into it. Nevertheless, a purge is still required. Husa[5] found the purge for flare stacks without a seal one-half the rate for unflared stacks. To cover the case of flame failure, he suggested the same purge rate for flares as for unflared vents. Thus, Fig. 6-2 may also be used for flares without a seal.

Light flammable gases, e.g., natural gas, are sometimes used for purge gas in flares. If a valve or pipe is opened, however, the natural draft created by the light gas can induct air, possibly forming a flammable mixture. Ignition by the pilots, then, produces an explosion hazard. Adequate design, operating, and maintenance procedures are necessary to prevent this possibility.

Unconfined Vapor-Cloud Explosions

The previous considerations in this section pertain to continuous or emergency emissions. Proper planning can minimize their fire and explosion hazards. Massive accidental releases of flammable vapors, however, are more difficult to cope with, and major disasters have resulted from ignition of such releases.* The Flixborough (England) Works explosion on June 1, 1974, occurred when a mammoth cyclohexane vapor cloud was ignited. It was equivalent to the force of 15 tons of TNT, killed 28 people, injured 89, and damaged 1821 houses.[12] That was not the first unconfined-vapor-cloud explosion, however. Strehlow[13] states that 108 accidental unconfined-vapor-cloud explosions have been documented over the past 42 years. The bulk of them were from heavier-than-air vapors that tend to stratify near the ground with less dispersion than light gases. The latter rise in the atmosphere as a result of buoyancy and generally do not produce large accumulations of gas at low levels. Moreover, vapor-cloud explosions usually result from leaks of flashing liquids, i.e., liquids under pressure at temperatures above their atmospheric boiling points, because of the enormous vapor volume that can be produced by these discharges.[9] Davenport[14] has indicated that the amount of material vaporized from flashing liquids is equal to the ratio of the superheat of the liquid material $c_p\ \Delta T$ to the heat of vaporization ΔH_v. For propane at 25°C and 946 kPa (boiling point at 101.325 kPa = −42.2°C)

$$\begin{aligned} c_p &= 2.412 \text{ kJ/kg·K} \\ \Delta T &= 67.2 \text{ K} \\ \Delta H_v &\approx 384.2 \text{ kJ/kg} \end{aligned}$$

Thus, (100)(2.412)(67.2)/384.2 = 42 percent of the liquid propane will immediately flash to air.

* *Vapor* and *gas* are used synonymously in this discussion of unconfined-vapor-cloud explosions. Atmospheric concentrations are for vapor releases from flashing liquids.

Atmospheric concentrations. Atmospheric-dispersion equations can be used to estimate the concentrations of flammable vapors.[15] For a ground-level point source

$$\chi(x,y,z) = \frac{100Q}{\pi\sigma_y\sigma_z u} \exp\left\{-\frac{1}{2}\left[\left(\frac{y}{\sigma_y}\right)^2 + \left(\frac{z}{\sigma_z}\right)^2\right]\right\} \quad (6\text{-}6)$$

Atmospheric-dispersion equations have been used by Burgess and Zabetakis[16] and Burgess et al.[17] to estimate concentrations of dense vapors from spills. Because heavy vapors tend to hug the ground, stable (E) atmospheric conditions are used in this text to approximate the spread of dense vapors under all light to moderate winds. (F stability was used by Burgess and Zabetakis[16] in calculating propane concentrations from a spill in complicated terrain. Turbulence created by plant structures, however, is likely to produce slightly better, but still relatively poor, dispersion, indicated by E stability.)

Standard deviations are given in Table 6-1.

Concentrations from Eq. (6-6) are for sampling periods of about 1 h. Momentary concentrations will be higher by a multiplier of around 2. (This factor of 2 is for stable atmospheric conditions. The factor of 5 used previously for elevated vents is to cover all stabilities.) Subsequent computations include this factor by multiplying Eq. (6-6) by 2. Also, for large releases calculated $\chi > 100$ percent at the source. Thus, a fictitious upwind source has been determined so that $\chi = 100$ percent at the true source of vapors.

Light winds give maximum concentrations. (Carried to an extreme, zero wind speed gives infinite χ. On the other hand, acceptance of that extreme assumption also requires that vapors do not travel, and $\chi = 0$ because of this effect.) Minimum credible wind speed for use in analysis of "worst cases" from accidental releases of dense vapors is probably near 1 m/s; meandering of a cloud at lower u can give more spreading and lower concentrations than with a sustained wind direction with $u \approx$ 1 m/s. Concentration contours for a vapor release at the ground of 100 mol/s with a wind of 1 m/s are shown in Fig. 6-5 (244 and 504 kg/min for propane and cyclohexane, respectively, for instance). The marked reduction in concentration with a wind of 5 m/s for otherwise identical conditions is shown in Fig. 6-6. (Average wind speeds for many locations are about 5 m/s.)

Large spills can give flammable concentrations for considerable down-

TABLE 6-1 Standard Deviations of Plume-Concentration Distribution for E Atmospheric Stability[15]

Downwind distance, m	σ_y, m	σ_z, m
<300	$0.0873x^{0.92}$	$0.0736x^{0.84}$
300–4000	$0.0873x^{0.92}$	$0.1771x^{0.69}$

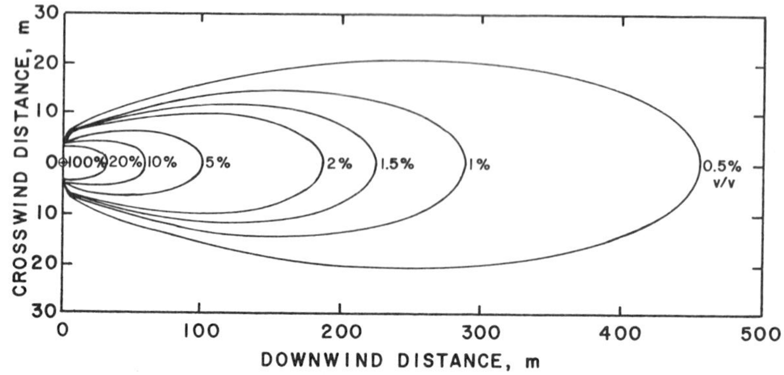

Fig. 6-5 Contours of estimated momentary concentrations (volume percent) at the ground from a continuous ground-level release of 100 mol/s of vapor. E stability with level terrain. Plan view with wind left to right. Emission point at $x = 0$, $y = 0$; $u = 1$ m/s. (Note that the crosswind scale is larger than the downwind scale.)

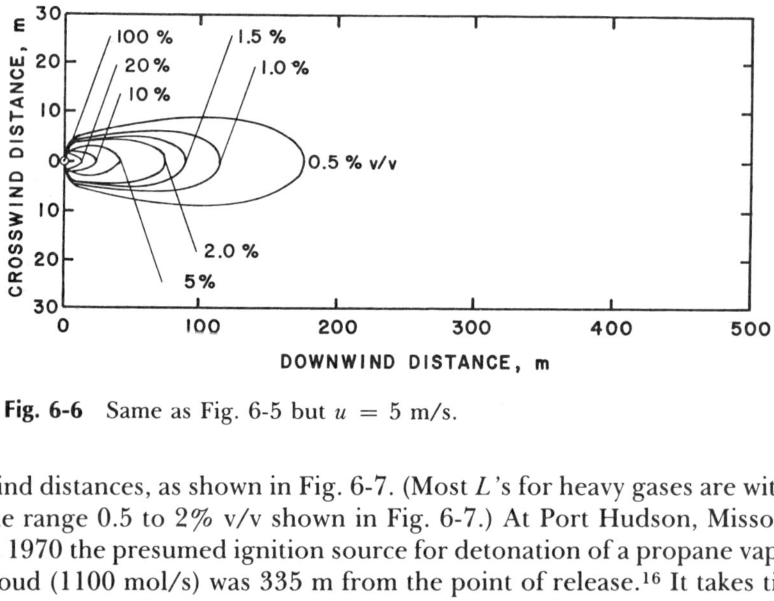

Fig. 6-6 Same as Fig. 6-5 but $u = 5$ m/s.

wind distances, as shown in Fig. 6-7. (Most L's for heavy gases are within the range 0.5 to 2% v/v shown in Fig. 6-7.) At Port Hudson, Missouri, in 1970 the presumed ignition source for detonation of a propane vapor-cloud (1100 mol/s) was 335 m from the point of release.[16] It takes time for vapors to disperse upward, however. Flammable concentrations from releases of dense vapors at the ground seldom extend to appreciable heights. Concentration isopleths at $z = 6$ m for a vapor release at the ground of 100 mol/s are shown in Fig. 6-8 ($u = 1$ m/s). (For the same conditions but with $u = 5$ m/s, $\chi < 0.37$ percent.) The maximum concentration at a given height can be determined by maximizing Eq. (6-6) with respect to x, as shown in Appendix C. The maximum heights to which specified concentrations extend are shown in Fig. 6-9.

The rate of release, not the total quantity emitted, is the primary

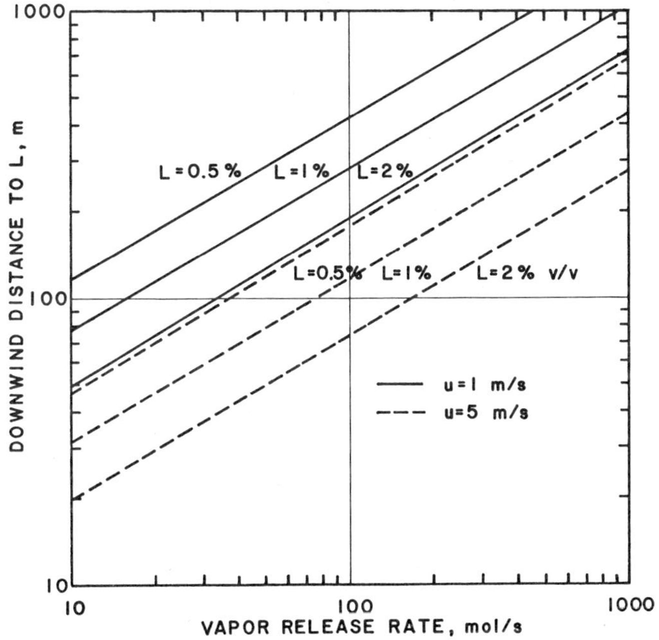

Fig. 6-7 Estimated downwind distance to lower flammability limit L % v/v at the ground in centerline of cloud vs. vapor-release rate. Continuous ground-level release from a point source with E stability over level terrain (momentary concentrations for L).

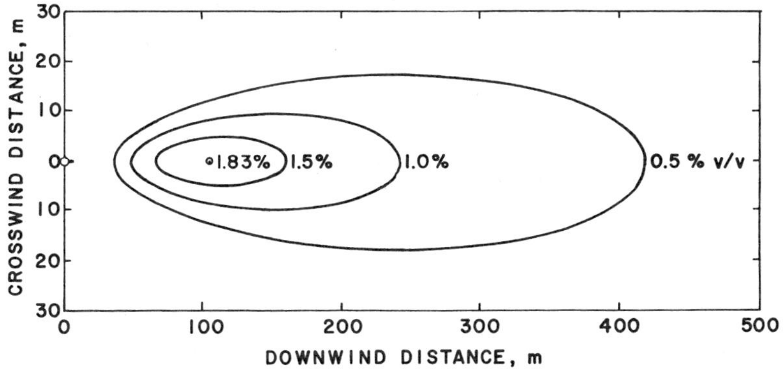

Fig. 6-8 Same as Fig. 6-5 but concentrations at a height z of 6 m.

criterion in assessing the explosion hazard potential of continuous releases. Once steady-state conditions are attained, concentrations remain the same no matter how much longer the release lasts. If an unconfined-vapor-cloud explosion does occur, vapor above U probably can take part in the explosion because of addition of air promoted by the incipient

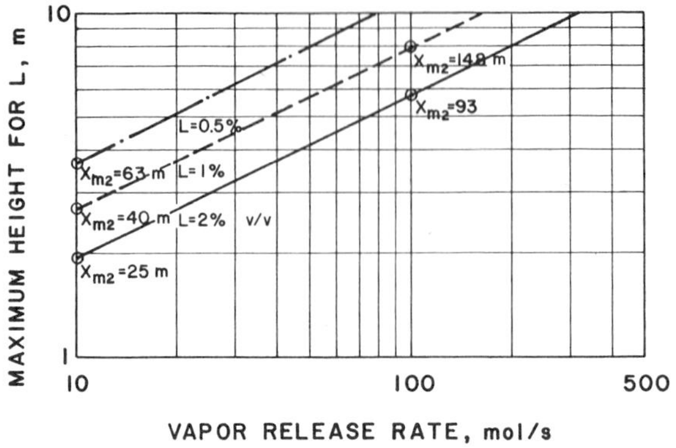

Fig. 6-9 Estimated maximum height for specified lower flammability limits L % v/v vs. vapor-release rate. Continuous ground-level release with E stability over level terrain from a point source of emission. (momentary concentrations for L). x_{m_2} is the downwind distance to maximum height for L for 10 and 100 mol/s of vapor; $u = 1$ m/s.

explosion. Burgess et al.[17] developed an expression for the volume of fuel between U and L. Their expression can be shortened to determine the volume of fuel between L and 100 percent, as follows:

$$V_f \approx \frac{Qx_L(b + d)}{u(b + d + 1)} \tag{6-7}$$

where b and d are the exponents on x in following general equations for σ_y and σ_z:

$$\sigma_y = ax^b$$
$$\sigma_z = cx^d$$

Then, from Eq. (6-6), with $y = z = 0$,

$$x_L = \left[\frac{(100)(Q)(2)}{\pi uLac}\right]^{1/(b + d)*} \tag{6-8}$$

With E stability for $x < 300$ m, $b = 0.92$ and $d = 0.84$ (Table 6-1). Therefore,

$$V_f \approx \frac{0.64Qx_L}{u} \tag{6-9}$$

Only the volume of fuel greater than L, that is, V_f, at the moment of ignition can take part in an explosion originating from a continuous release. In these circumstances, the total quantity emitted is irrelevant.

*V_f in Eq. (6-7) is not multiplied by 2 since the amount of gas greater than L at a given time should not be changed by meandering of the cloud with accompanying higher momentary concentrations at a given location. Nevertheless, Q in Eq. (6-8) is multiplied by 2 to determine the maximum downwind distance to momentary concentrations equal to L.

Nevertheless, total quantity is the key item with an explosively dispersed vapor cloud (puff). A vapor release is essentially steady, not instantaneous, if it lasts longer than σ_y/u.[17] Equations are available to determine the downwind concentrations from a puff, though dispersion parameters for such instantaneous sources are uncertain.[15] A puff expands as it travels downwind, akin to a balloon being filled, with concentrations decreasing all the while. Nevertheless, with large instantaneous releases, high concentrations of fuel can occur at extraordinarily large distances from the release point. Because of the brevity of the emission, however, these high concentrations are short-lived at any fixed location; if the distance between L's on the downwind axis of a cloud is 100 m at a given time, for instance, the portion of the cloud with $\chi > L$ will last for only about 1 min at the then location of the cloud with $u = 2$ m/s. Davenport[14] indicates that his data suggest that explosively dispersed clouds do not represent the most severe vapor-cloud explosion condition; instead vapor-cloud explosions from continuous emissions have caused the peak losses.

Pressure. If a vapor cloud is ignited, a vapor-cloud explosion does not necessarily result; only a fireball, bad enough in itself, accompanied by a nondamaging "whoosh" may occur. Deflagration does not give significant overpressure.[18] With a sufficiently powerful ignition source, detonation may occur in the open.[19] Lee et al.[18] have indicated, however, that natural ignition sources are usually too weak to cause detonations in most fuel-air mixtures. Thus, Strehlow and Baker[20] conclude that detonative combustion must always occur before a destructive blast wave is produced. Gugan[21] has developed a theory for unconfined vapor-cloud explosions in which "pinch"-induced flame acceleration occurs for flammable concentrations of dense vapors in air. He believes detonation need not occur for generation of an unconfined vapor-cloud explosion. Whether or not detonation must occur to produce blast pressure is a point that will not be argued here. Suffice it to say that high and damaging blast pressures occur. Davenport[22] has indicated that confinement and subsequent ignition of a portion of the vapor cloud produces a high-energy igniter for the remainder of the cloud; Watts[23] has shown that the yield of ethylene-air explosions in polyethylene chambers increased 150 times with slight confinement. Thus, opportunities for confinement of any portion of a vapor cloud should be eliminated to the extent feasible. This can be accomplished in part, for example, by sealing up spaces to prevent vapor entry. In addition to the apparent need for powerful igniters to produce damaging overpressure, unconfined-vapor-cloud explosions also appear to depend on the size of the cloud; large clouds are required, although the minimum required size has not been determined with precision.

Procedures to deduce blast pressures and effects from vapor-cloud explosions are in the embryonic stage; considerably more investigation is required before the phenomenon is understood sufficiently. The common method is to calculate a mass of TNT equivalent to the fuel in the vapor cloud. Relative to a concentrated solid-explosive charge, however, a vapor-cloud explosion is nonideal; the latter is many times larger than the former and a vapor cloud has a complex pattern of fuel-air concentrations. Also, Gugan's development implies pressure durations one to two orders of magnitude longer than for conventional explosives releasing similar energies, i.e., sufficiently long to be considered static pressure.[21] Moreover, Strehlow and Baker[20] have cautioned that "the concept of 'TNT equivalence,' which is widely used in safety studies, is also very inexact and may be quite misleading. But, this concept will undoubtedly be used to estimate 'yields' of accidental explosions until better measures are available." The blast pressure from an unconfined vapor-cloud explosion of 100 mol/s of propane for conditions described in Fig. 6-5 are discussed in that context in the following example.

Example

$$Q = (100)(22.4146 \times 10^{-3})\ 298.15/273.15 = 2.447\ \text{m}^3/\text{s}$$
$$L = 2.2\%\ \text{v/v}$$

From Eqs. (6-8) and (6-6) for $x < 300$ m with peak $L = 2.2\%$ v/v, $x_L =$ 175 m from the true source of gas (also, from Fig. 6-7 extrapolated and Fig. 6-5.) Thus, from Eq. (6-9)

$$V_f \approx \frac{(0.64)(2.447)(175)}{1} = 274\ \text{m}^3$$

Therefore,

$$W_c \approx 493\ \text{kg propane}$$

As discussed by several researchers, e.g., Brasie and Simpson[24] and Strehlow and Baker,[20]

$$W_{\text{TNT}} = -\frac{\alpha\,\Delta H_c W_c}{4.52}$$

where 4.52 is the heat of explosion of TNT in megajoules per kilogram. The heat of combustion ΔH_c of propane is -46.35 MJ/kg. The maximum probable "yield" α is difficult to quantify. Based on the volume of fuel greater than L, that is, V_f, maximum $\alpha \approx 0.2$. (Note that this volume is not the total quantity emitted for a continuous release.) As will be evident shortly, however, overpressure at a specific location is influenced more by R and by the placement of the center of the cloud than by W_c. Thus, W_{TNT} for $\alpha = 0.2$ and for $\alpha = 0.1$ is 1010 and 505 kg, respectively.

The placement of the center of a potential blast is uncertain, but for a surface explosion it seems reasonable to assume the center to be where the initial flammable vapor cloud has its greatest depth, wherever the ignition source is. (The reported center of the blast at Port Hudson, Missouri, was 295 m from the point of gas release and 110 m from the presumed point of ignition.[16]) Thus, in the manner of Appendix C the horizontal distance from the true source to the maximum height for peak $L = 2.2$ percent is 87 m ($z = 5.5$ m). This is close to the 118-m distance to the stoichiometric concentration in the centerline of the cloud at the ground. (The crosswind distance to L from the centerline of the cloud at the ground is about 10 m at $x = 87$ m.) Estimated p_{so}, p_r, i_s, i_r, and t_0 outside the cloud at R from this assumed center from Fig. 4-9 are given in Table 6-2.

Within the cloud appreciably higher pressures are not impossible. It is essential to recognize that the epicenter of a vapor-cloud explosion is not necessarily at the vapor-release point; peak pressures may be considerably removed from the origin of the vapors because of the previously described lateral extent of a cloud, particularly during poor atmospheric-dispersion conditions.

Prevention and protection. The wind speed in the Flixborough disaster averaged about 6 m/s, with gusts to 10 m/s. Thus, these favorable atmospheric-dispersion conditions could not prevent a vapor-cloud explosion from the massive release of cyclohexane that occurred. Accordingly, *the primary defense against vapor-cloud explosions has to be prevention of an accidental release in the first place.* Excess-flow valves should be considered as a means of preventing large emissions of flammable vapors and gases. (An *excess-flow valve* is a device in which the flow of material is stopped by the valve's closing when the flow exceeds a predetermined amount. The valve closes as a result of the pressure drop across the valve caused by an excessive flow of liquid or gas.) Alternatively or in addition a remotely operated shutoff valve can be used to arrest the release of liquid or gas from a leak downstream of the valve. This can be operated manually from a number of locations. Also, there is growing use of combustible-gas analyzers to warn of a leak; the shutoff valve can be tied into an array of these analyzers so that the valve is shut off if any analyzer indicates a predetermined concentration. Nevertheless, it must be realized that there is very little time to react if a release does occur. Vapors will travel 0.5

TABLE 6-2

R, m	Z, m/kg$^{1/3}$	p_{so}, kPa	p_r, kPa	i_s, kPa·ms	i_r, kPa·ms	t_0, ms
			$W_{TNT} = 1010$ kg			
25	2.49	180	600	1324	3913	17.5
50	4.98	44	100	672	1555	33
100	9.96	15	31	321	742	47
150	14.95	9	18	221	502	54
			$W_{TNT} = 505$ kg			
25	3.14	100	290	860	2229	19
50	6.28	30	66	414	955	30
100	12.56	11	24	207	477	40
150	18.84	6.5	13	143	318	45

km, for example, in about 4 and 1.5 min with wind speeds of 2 and 5 m/s, respectively.

Although the initial design of a plant may incorporate ample safety features to guard against accidental emissions, plant modifications might introduce traps that could lead to potentially disastrous releases. *Every modification, even if seemingly minor, should be checked* to assure that new hazards have not been introduced, that the modification is consistent with the design of the original installation, including materials of construction, and that applicable standards have been complied with.

Protection measures may be necessary, even if prevention procedures and equipment have been used. The question inevitably arises whether a vapor cloud should be ignited intentionally to avoid a vapor-cloud explosion. Intentional igniters should not be installed without a thorough analysis of the ramifications of intentional ignition; in the event of an accidental release, a fire is inevitable and delayed ignition could produce a disasterous vapor-cloud explosion. In general, purposeful ignition has not been practiced on an important scale.

Water sprays have been used to reduce the hazards of unconfined-vapor-cloud explosions. Watts[23] has shown that the most important function of the water spray is dilution of the gas by induced air. Nevertheless, it is not likely that water sprays can be effective in sufficient dilution of *large releases.* On the other hand, large fans could be considered as protection against vapor-cloud explosions because of the enormous air dilution they can produce. An array of, say, four fans, each with a flow of 500 m^3/s directed straight up, could reduce the threat of an explosion. Strategically placed, they probably would influence air motion during light and worst wind conditions so as to induct vapors. (Air coolers can do the same thing, to a degree, and judicious placement of them in initial plant design can serve a useful secondary purpose of dilution of flammable vapors from an accidental release.) The four fans could dilute 285×10^3 kg/h propane to its L. Moreover, ignition sources at elevated locations are fewer than at the ground, and there would be further enhanced dilution of vapors by dispersion from the elevated position. Operation of the fans could be triggered by a set of combustible-gas analyzers.

The design of control rooms and other buildings requires assessment to provide protection against the potential destructive effects of vapor-cloud explosions. The Manufacturing Chemists' Association (MCA) (now the Chemical Manufacturers' Association, CMA) has provided guidelines on new control houses in its Safety Guide S22, Siting and Construction of New Control Houses for Chemical Manufacturing Plants. For vapor-cloud explosions, an explosion hazard exists if a flammable vapor release of more than 9.1 Mg in a 5-min period is possible (MCA category C). A control house in this category should be located a minimum of 30 m away from the flammable-material-containing equipment or highly exothermic reactions. Several other criteria are specified. Walls should be designed for a peak reflected pressure of 172 kPa and a duration of 20 ms. Where design for a vapor-cloud deflagration is more applicable, each wall should be designed for a peak reflected pressure of 30 kPa and a positive duration of 100 ms.

Additional design factors to consider are discussed in the following paragraphs.

Buildings that house personnel should be removed as far as practicable from the potential sources of vapors; congregation of occupied buildings close to processing areas increases the risk of damage and injury from a vapor-cloud explosion. Moreover, the possibility of locating control rooms and other buildings upwind of the prevailing wind direction from processing areas should be considered. The prevailing wind (most frequently observed direction) for Philadelphia is from the west southwest 11 percent of the time for a 16-point compass. Winds blow out of the southwest-northwest quadrant 40 percent of the time and out of the northeast-southeast quadrant only 19 percent of the time. Thus, placement of these buildings generally west of processing areas in midlatitudes can reduce risk. Marshall[25] has emphasized the following additional five points for location and construction of control rooms:

> The first concerns materials of construction. There is much evidence that when an explosion strikes, the brick panels of concrete-framed buildings cave in. Even worse, the bricks themselves can become missiles. Under conditions of nuclear attack, reinforced concrete buildings have shown remarkable resistance. In Hiroshima, for example, a reinforced concrete-and-steel building—designed to be earthquake resistant—contained no brick panels and stood up well even though it was only 300 yd from the epicenter.
>
> At Flixborough, the reinforced concrete walls of the sulfate store—which was about 150 yd from the believed epicenter—cracked but did not collapse even though the walls were buttressed to withstand pressure from inside and not from outside. Yet, the sulfuric acid control room—about the same distance away—had its brick walls blown in.
>
> The second point is that a building should preferably have only one floor. If it must have two or more, on no account should the top floors have heavy equipment on them. The main Flixborough control room, in which 18 lives were lost [*sic*], had electrical switch gear on the floor above the room housing the control panel.
>
> The third point concerns control-room siting. This room should have an open line of retreat, because the personnel within should be the last ones to leave a disaster area. A central situation will not allow this.
>
> Fourth, a control room should be so sited that other structures will not fall on it, as occurred at Flixborough, where the main pipe bridge collapsed on top of the control room.
>
> Fifth, a control room must have forced ventilation and emergency supplies, as well as small safety-glass windows. Provisions must also be made so that control rooms have disaster-proof data storage to permit an investigation, should disaster strike. It is also a good idea for plant operators to practice with mock-up disaster situations.

Furthermore, because of the propensity of heavy gases to hug the ground, air intakes for operating buildings should be elevated to reduce chances of drawing in flammable concentrations. An air intake at 6 m, for example, will provide much better protection than one near the ground (see Fig. 6-9).

References

1. Grumer, J., M. E. Harris, and V. R. Rowe, "Fundamental Flashback, Blowoff, and Yellow-Tip Limits of Fuel Gas-Air Mixtures," *U.S. Bur. Mines Rep. Invest.* 5225, July 1956.

2. Grumer, J., "Flashback and Blowoff Limits of Unpiloted Turbulent Flames," *Jet. Propul.,* vol. 28, no. 11 pp. 756–758, November 1958.

3. Khitrin, L. N., P. B. Moin, D. B. Smirnov, and V. U. Shevchuk, "Peculiarities of Laminar- and Turbulent-Flame Flashbacks," pp. 1285–1291 in *10th Int. Symp. Combust.,* Combustion Institute, Pittsburgh, 1965.

4. Litchfield, E. L., Flame Arresters, *ISA Monogr.* 110, pp. 75–84, 1965.

5. Husa, H. W., "How to Compute Safe Purge Rates," *Hydrocarbon Process. Pet. Refiner,* vol. 43, no. 5, pp. 179–182, May 1964.

6. Bodurtha, F. T., "The Behavior of Dense Stack Gases," *J. Air Pollut. Control Assoc.,* vol. 11, no. 9, pp. 431–437, September 1961.

7. Hoot, T. G., and R. N. Meroney, "The Behavior of Negatively Buoyant Stack Gases," *pap. presented at the 67th Annu. Meet. Air Pollut. Control Assoc., Denver, June, 1974.*

8. Guide for Pressure Relief and Depressuring Systems, *Am. Pet. Inst. RP 521,* September 1969.

9. Kletz, T. A., "Unconfined Vapour Cloud Explosions," *Chem. Eng. Prog. 11th Loss Prev. Symp., Houston, 1977,* pp. 50–58.

10. American National Standards Institute, Power Piping, *ANSI* B31.1, The American Society of Mechanical Engineers, New York, 1977, and addenda.

11. Reider, R., H. J. Otway, and H. T. Knight, "An Unconfined Large-Volume Hydrogen/Air Explosion," *Pyrodynamics,* vol. 2, no. 4, pp. 249–261, May 1965.

12. Warner, Sir Frederick, "The Flixborough Disaster," *Chem. Eng. Prog.,* vol. 71, no. 9, pp. 77–84, September 1975.

13. Strehlow, R. A., Unconfined Vapor-Cloud Explosions: An Overview, pp. 1189–1200 in *14th Int. Symp. Combust.* Combustion Institute, Pittsburgh, 1973.

14. Davenport, J. A., "A Survey of Vapor Cloud Incidents," *Chem. Eng. Prog.,* vol. 73, no. 9, pp. 55–63, September 1977.

15. Turner, D. B., Workbook of Atmospheric Dispersion Estimates, *Environ. Prot. Agency (U.S.) Publ.* AP-26, rev., 1970.

16. Burgess, D. S., and M. G. Zabetakis, Detonation of a Flammable Cloud Following a Propane Pipeline Break, *U.S. Bur. Mines Rep. Invest.* 7752, 1973.

17. Burgess, D., J. N. Murphy, M. G. Zabetakis, and H. E. Perlee, "Volume of Flammable Mixture Resulting from the Atmospheric Dispersion of a Leak or

Spill," pp. 289–303 in *15th Int. Symp. Combust.* Combustion Institute, Pittsburgh, 1974.

18. Lee, J. H., C. M. Guirao, K. W. Chiu, and G. G. Bach, "Blast Effects from Vapor Cloud Explosions," *Chem. Eng. Prog. 11th Loss Prev. Symp., Houston, 1977,* pp. 59–70.

19. Zabetakis, M. G., "Flammability Characteristics of Combustible Gases and Vapors," *U.S. Bur. Mines Bull.* 627 (*USNTIS* AD 701 576), 1965.

20. Strehlow, R. A., and W. E. Baker, "The Characterization and Evaluation of Accidental Explosions," *Prog. Energy Combust. Sci.,* vol. 2, no. 1, pp. 27–60, 1976.

21. Gugan, K., *Unconfined Vapor Cloud Explosions,* Gulf Publishing, Houston, Tex., 1979.

22. Davenport, J. A., "Prevent Vapor Cloud Explosions," *Hydrocarbon Process.,* vol. 56, no. 3, pp. 205–214, March 1977.

23. Watts, J. W., Jr., "Effects of Water Spray on Unconfined Flammable Gas," *Chem. Eng. Prog., 10th Loss Prev. Symp., Kansas City, 1976,* pp. 48–52.

24. Brasie, W. C., and D. W. Simpson, "Guidelines for Estimating Damage Explosion," *Chem. Eng. Prog. 2d Loss Prev. Symp., St. Louis, 1968,* pp. 91–102.

25. Marshall, V. C., "Process-Plant Safety: A Strategic Approach," *Chem. Eng.,* vol. 82, no. 27, pp. 58–60, Dec. 22, 1975.

7

Hazardous Compounds, Reactions, and Operations

Hazardous compounds and reactions are many and varied. It is important to understand the types of compounds and reactions that may cause hazardous conditions. Certain operations can also create risks that require correction. Some hazardous compounds, reactions, and operations are discussed in this chapter.

7-1 Vapor Explosions

When a hot liquid comes into contact with a cold liquid, sudden vaporization from the resulting superheated cold liquid can produce a shock wave. This is a physical phenomenon, often called a *vapor explosion* (not to be confused with a vapor-cloud explosion). Sometimes these explosions are unexpected, but they can and do cause damage and injury.[1] The energy source is the sensible heat plus the heat of fusion of the hot liquid; various theories about the cause of these vapor explosions are covered in Refs. 1 to 7. Genco and Lemmon[2] estimated that the peak overpressure 3 m from a spill of 45 kg of slag at 1482°C on a puddle of water 0.56 m^2 in area and 1.59 mm deep could be 33 kPa.

The cold liquid is generally water, and the hot liquid is often molten metal. In addition, molten salts can produce vapor explosions if they are quenched or come into contact with water. The usual mode of contact is for the molten material to be spilled on water, but vapor explosions can also occur when water is poured onto the melt.

Water, may act as a hot liquid if it comes into contact with a relatively cold liquid, such as chlorodifluoromethane, which boils at −41°C. Vapor explosions can occur, if liquid ethane containing a small percentage of heavier hydrocarbons is spilled on water.[3,4]

Another type of vapor explosion may occur when a hot pressurized liquid is suddenly depressurized. The bulk liquid may undergo a spontaneous nucleation with sudden vaporization and development of a shock wave. More research is needed on this topic to delineate the mechanism involved.

7-2 Hazardous Compounds and Reactions

Sources of Information

Comprehensive compilations of hazardous reactions are given in Refs. 8 to 10. The Manufacturing Chemists' Association (now the Chemical Manufacturers' Association) has published four volumes of case histories of accidents in the chemical industry.[11–14] A classified index is included to assist readers searching for case histories involving specific chemicals, specific pieces of equipment, or specific occupations.

Sources of safety information for an alphabetical list of 503 chemicals are provided in appendix 2 of Ref. 15. A list of 265 explosions with literature references is given in appendix 3 of the same publication. Explosions involving the following 23 categories are covered in the list:

1. Acids
2. Acid anhydrides
3. Acid halides
4. Alcohols
5. Aldehydes
6. Amides
7. Amines
8. Azo and diazo compounds
9. Esters
10. Ethers
11. Halogen compounds
12. Hydrocarbons
13. Ketones
14. Nitrates
15. Nitriles
16. Nitro compounds
17. Nitroso compounds
18. Peracids and perchlorates
19. Peroxides
20. Phenols
21. Sulfonyl chlorides
22. Unsaturated compounds
23. Miscellaneous causes

Data on general classes of chemicals are presented in Data Sheet 7-23S, Loss Prevention Data, of the Factory Mutual Engineering Corp. Their Data Sheets 7-46 and 17-11 outline hazards and recommended correction procedures for chemical reactors and reactions. The Chemical Hazards Response Information System (CHRIS) manual, an official publication of the U.S. Coast Guard, consists of the following four volumes, which can be obtained from the Superintendent of Documents, U.S. Government Printing Office:

CG-446-1 A Condensed Guide to Chemical Hazards
CG-446-2 Hazardous Chemical Data
CG-446-3 Hazard Assessment Handbook
CG-446-4 Response Methods Handbook and Appendix

In addition, a myriad of reactions, including hazardous ones, are discussed in Mellor[16] and Kirk and Othmer.[17] Finally, document and data banks of such organizations as Chemical Abstracts Service, Engineering Index, Inc., and, for United States government reports and announcements, The National Technical Information Service are available for searching.

The foregoing list is not all-inclusive but does contain information that should be useful in explosion prevention and protection practice.

Hazardous Compounds

Heat evolution, i.e., exothermicity, is ordinarily the essential requirement for a chemically hazardous compound; such materials can release potentially destructive energy upon introduction of sufficient activation energy. Compounds with double or triple bonds are particularly likely to decompose with potentially damaging liberation of energy. Also, *many compounds with nitrogen are unstable,* and special caution is prudent with them.

Endothermic compounds. Compounds having positive heats of formation are called endothermic compounds; energy input is required for their formation, and that energy is realizable if the compound decomposes. (ΔH_f is not directly related to the sensitivity of a compound to decomposition or to the rate of decomposition.) Heats of formation for many compounds are given in handbooks of chemistry and in the *JANAF Thermochemical Tables.*[18] The heat of formation of a combustible compound can be calculated from its heat of combustion, as follows.

Example

The heat content of each element in its standard state, 25°C, is zero. (The sign of heat evolution is negative.) Thus, for a $C_nH_xO_yN_z$ compound

$$C_nH_xO_yN_z + \left(n + \frac{x}{4} - \frac{y}{2}\right) O_2 \rightarrow nCO_2 + \frac{x}{2} H_2O(l) + \frac{z}{2} N_2$$

ΔH_c equals the sum of the heats of formation of the products of a reaction minus the sum of the heats of formation of the reactants. Therefore,

$$\begin{aligned} \Delta H_f(C_nH_xO_yN_z) &= -\ \Delta H_c + n\ \Delta H_f(CO_2) + 0.5x\ \Delta H_f[H_2O\ (l)] \\ \Delta H_f(CO_2) &= -393.8\ \text{kJ/mol} \\ \Delta H_f[H_2O\ (l)] &= -286.0\ \text{kJ/mol} \end{aligned}$$

Thus,

$$\Delta H_f(C_nH_xO_yN_z) = -\Delta H_c - 393.8n - 143.0x$$

For benzoic acid ($C_7H_6O_2$), for example,

$$C_7H_6O_2 + 7.5O_2 \rightarrow 7CO_2 + 3H_2O(l) \qquad \Delta H_c = -3232 \text{ kJ/mol}$$

where

$$n = 7$$
$$x = 6$$
$$y = 2$$
$$z = 0$$

$$\Delta H_f(C_7H_6O_2) = 3232 - (393.8)(7) - (143.0)(6)$$
$$= -382.6 \text{ kJ/mol}$$

Benzoic acid is stable, having a negative heat of formation.
For acetylene (C_2H_2)

$$C_2H_2 + 2.5O_2 \rightarrow 2CO_2 + H_2O(l) \qquad \Delta H_c = -1300 \text{ kJ/mol}$$

$$n = 2$$
$$x = 2$$
$$y = 0$$
$$z = 0$$

$$H_f(C_2H_2) = 1300 - (393.8)(2) - (143.0)(2)$$
$$= 227 \text{ kJ/mol}$$

Acetylene is an endothermic compound and can decompose explosively. Chlorine dioxide (ClO_2), with a heat of formation of 103 kJ/mol, is another endothermic compound.

In addition, the heats of formation and other thermodynamic data of compounds can be estimated from their molecular structure by CHETAH.*

Nitrogen compounds with positive heats of formation include azides (—N=N≡N), diazo compounds (—N=N—), diazonium salts (—N≡N±), and ring structures containing several nitrogen atoms when some have double bonds. The importance of double and triple bonds between nitrogen atoms in causing instability is evident from this list. Some of these compounds are also sensitive to decomposition.

Fuel plus oxidizer. Compounds with fuel and oxidizer together in the same compound are prone to decomposition and are one of the leading causes of industrial fires and explosions. They include organic nitrates (—O—N$\lessgtr$$^{O}_{O}$) and nitro compounds (—N$\lessgtr$$^{O}_{O}$), organic perchlorates (—C1O_4), chlorates (—C1O_3), and chlorites (—C1O_2). Ammonium

*CHETAH is the ASTM Chemical Thermodynamic and Energy Release Evaluation Program.[19,20] In addition to estimating thermodynamic properties, CHETAH uses the thermodynamic properties and other analysis to make energy-hazard evaluations.

compounds with these oxidizers may decompose explosively. The nearer the compound approaches stoichiometry, the higher the energy of decomposition becomes. Moreover, the presence of easily oxidized substituents such as amino, hydroxyl, or methyl groups on the aromatic nucleus lowers the thermal stability of aromatic nitro compounds as measured by the energy of activation and the temperature required for heat evolution of 14 mJ/(s)(kg.)[21]

The ΔH_f for compounds containing fuel plus oxidizer is not necessarily positive. Heat of decomposition ΔH_d may be evolved from compounds possessing negative heats of formation because the products of the decomposition have still lower heats of formation. As an example, consider the decomposition of perchloric acid ($HClO_4$)

$$HClO_4(l) \rightarrow 0.5H_2O(l) + 0.5Cl_2(g) + 1.75O_2(g)$$

The heats of formation of $HClO_4(l)$ and $H_2O(l)$ are -40.6 and -286.0 kJ/mol, respectively. Thus,

$$\Delta H_d = (0.5)(-286.0) + 0 + 0 - (-40.6) = -102.4 \text{ kJ/mol}$$

CHETAH can also be used to determine ΔH_d. It selects the products of decomposition so as to maximize ΔH_d.

Peroxy compounds. Since the oxygen linkage in peroxy compounds (—O—O—) is easy to rupture, these compounds are inherently unstable. When handled as prescribed by manufacturers, however, commercially available organic peroxides can be used safely. A bulletin on Suggested Relative Hazard Classification of Organic Peroxides has been published by The Society of the Plastics Industry, Inc. Test methods and the classification system are described in the bulletin. Dilution of the peroxide with a suitable solvent is the most common method of reducing its explosion hazard.[22] Limiting the quantity in a single container is another way of minimizing the risk. Development, stabilization, and uses of organic peroxides are discussed in nine articles in *Industrial and Engineering Chemistry* and *I&EC Product Research and Development,* December 1964.

Other decomposable peroxy compounds and examples of them are alkyl hydroperoxides (*t*-butyl hydroperoxide, $C_4H_9—O—O—H$), peroxyesters (t-butyl peroxyacetate, $C_4H_9—O—O—C(=O)CH_3$), and peroxyacids (peracetic acid, $CH_3C(=O)O—OH$). Peracetic acid is a hazardous compound because in sufficiently high concentrations it can detonate in either the vapor or liquid phase.[23]

Nonvolatile peroxides can form if diethyl ether is left standing for an appreciable time in contact with air. They may explode in laboratory or plant operations if the ether is evaporated and the peroxide residue is overheated. The control of peroxidizable compounds is discussed by Jackson et al.,[24] who cover the following topics:

1. Structure of peroxidizable compounds
2. Examples of peroxidizable compounds
3. Handling procedures (labeling, storage, and inventory)
4. Distillation and evaporation of peroxidizable compounds
5. Detection of peroxides

Hazardous Reactions

Unexpected explosions often occur when fuel and oxidizer are mixed or where heat is otherwise released. For instance, chlorine (oxidizer) and grease (fuel) can explode. Particular caution for these reactions is required, whether they are on a small scale or large scale; the possibility of unusual reactions occurring to give heat release should be investigated. CHETAH can be used to estimate the maximum amount of energy available from a reaction mixture. In addition, tests may be required.

Although several types of reactions are liable to be hazardous, oxidations, nitrations, chlorinations, and polymerizations cause most plant reactor hazards. (Oxidations are discussed further below. Chlorinations are covered in Refs. 25 and 26.) These are all exothermic reactions. Close control of temperature and agitation are essential. In general, some common safety principles are applicable to all of them:

1. Inert diluents to serve as a heat sink should be used to the maximum extent feasible. An explosion is highly unlikely if water exceeds 55 wt% in the initial solution.

2. Continuous operations are often less hazardous than batch processes. With continuous feeds, the *dangerous component must not be allowed to accumulate;* reaction temperature needs to be reached and catalyst present before feed is started.[27] Consideration should be given to stopping the feed of the dangerous components automatically if reaction temperature in continuous processes goes below the set limit.

A wide variety of oxidizers are used to dehydrogenate or to introduce oxygen into organic compounds. They include permanganates; inorganic chlorine compounds with positive chlorine valence, such as sodium chlorate, $NaClO_3$; peroxides; nitric acid; and air. Oxidations are performed in the liquid phase and in the vapor phase. Oxidations by hydrogen peroxide and air are considered further in the following paragraphs.

Hydrogen peroxide (H_2O_2) vapors over water–hydrogen peroxide solutions are not flammable at atmospheric pressure (101.325 kPa) below 26% v/v H_2O_2. This corresponds to equilibrium conditions over 74 wt% H_2O_2 in water at the boiling point of the mixture at 101.325 kPa and 128°C. At higher concentrations detonation of H_2O_2 vapor can occur.[28] Also, if they contain less than about 20% v/v H_2O_2, vapor mixtures of H_2O_2 with water and/or organic materials are not flammable at 101.325 kPa abs whatever the concentration of the organic.[29] Since 1 mol H_2O_2

yields 0.5 mol O_2, this 20% v/v H_2O_2 is approximately equivalent to the MOCs in Table 2-4.

Detonation of H_2O_2–H_2O mixtures in the liquid phase is hard to accomplish; only solutions that are nearly anhydrous (99 wt% H_2O_2) have been observed to detonate.[29] Nevertheless, some mixtures of H_2O_2 and combustible materials can detonate if subjected to a strong mechanical or explosive shock. Furthermore, H_2O_2 and other peroxy compounds can decompose to evolve oxygen, which then can form flammable mixtures with combustible vapor. The presence of high oxygen concentration widens the flammable range, markedly decreases the minimum ignition energy compared with air, and produces much higher explosion pressure and rate of explosion-pressure rise compared with air. A nitrogen purge to keep O_2 below the MOC is one way to provide nonflammable conditions in the vapor. Oxygen-monitoring guidance is given in Chap. 2.

Solutions of H_2O_2 with water and/or organic material in any amount do not explode when subjected to strong mechanical or explosive shock if H_2O_2 is less than about 30 wt% in those mixtures.[30] More H_2O_2 may be permissible without compromising safety for some organic materials, but such higher concentrations should be determined by testing. Note that immiscible systems can explode at interfaces and that increased dispersion may cause an immiscible system to be as dangerous as a miscible system.[30] Possible changes in a system should also be studied to determine whether an ostensibly safe system can be rendered hazardous by such changes. In addition to the 30% H_2O_2 criterion, an explosion is improbable if the numerical value of the energy released is not over 1.88 MJ per kilogram of reacting solution[30,31] (all heat sources and sinks considered). An example with ethanol is given below.[30]

Example

With a 100-kg solution containing 38 wt% H_2O_2, 18% C_2H_5OH, and 44% H_2O, 9.4 kg ethanol is in excess over the amount to react with the 38 kg H_2O_2, as determined from

$$C_2H_5OH(l) + 6H_2O_2(l) \rightarrow 2CO_2(g) + 9H_2O(g)$$

$\Delta H = -6.36$ MJ/kg for the stoichiometric mixture. Heats of vaporization of ethanol and water are 0.866 and 2.26 MJ/kg, respectively.

$$\Delta H = \frac{-(6.36)(8.6+38) + (9.4)(0.866) + (44)(2.26)}{100}$$

$$= -1.88 \text{ MJ/kg}$$

This solution can become explosible if sufficient ethanol evaporates. In peroxide-rich solutions, the heat of decomposition of excess H_2O_2, -2.89 MJ/kg, and the positive heats of vaporization of the H_2O formed (18/34 H_2O_2) plus the original H_2O are added to the heat of reaction.

Air is used as an oxidizing medium in both the liquid and the vapor phase. In the liquid phase, these oxidations include acetaldehyde to acetic acid, cyclohexane to a mixture of cyclohexanone and cyclohexanol, and paraxylene to dimethyl terephthalate. Air is used to dehydrogenate methanol to formaldehyde and to oxidize naphthalene to phthalic anhydride in the vapor phase, for example. In liquid-phase air oxidations, explosion safety ordinarily depends upon the dearth of oxygen in off-gases. Oxygen should be monitored to assure nonflammable conditions by virtue of deficient oxygen. Also, loss of normal reaction temperature can be used as a signal for possible breakthrough of air to give high and possibly unsafe concentrations of oxygen.

In vapor-phase oxidations with air, the fuel in the feed is usually above the upper flammability limit. Operations slightly below U have been performed, however, without ignition by the hot catalyst with ample gas flow to prevent flashback.[32] Explosion protection, such as explosion venting, may also be necessary for the reactor because of abnormal flammable concentrations that may develop during startups or upsets.

Thermal Explosions

Thermal explosions can occur when hazardous materials decompose exothermically and self-heat to produce a prodigious rate of reaction. The rates of heat generation by chemical reaction and heat loss in a fluid are[27,33,34]

$$\text{Heat generation} = aZ\,\Delta H \exp\left(\frac{-E_a}{R_u T_R}\right) \tag{7-1}$$

$$\text{Heat loss} = U'A''(T_R - T_A) \tag{7-2}$$

where Z is the Arrhenius preexponential factor and ΔH is in kilojoules per mole.

If the heat loss from a reaction is more than the heat generation, the reaction will subside. On the other hand, thermal explosions can occur when heat generation cannot be dissipated adequately. An induction period is necessary to give a thermal explosion. If conditions are adiabatic, the time to explosion (maximum rate of decomposition) τ_{ad} in seconds can be determined with the help of Eq. (7-1). Thus,

$$(a)(\text{MW})(c_p)\,\frac{dT_R}{dt} \times 10^{-3} = aZ\,\Delta H \exp\left(\frac{-E_a}{R_u T_R}\right) \tag{7-3}$$

and[34,35]

$$\tau_{ad} = \frac{(c_p)(\text{MW}) \times 10^{-3}}{Z\,\Delta H}\,\frac{R_u T_R^2}{E_a}\,\exp\frac{E_a}{R_u T_R} \tag{7-4}$$

Times to adiabatic explosion are shown in Fig. 7-1.

Even if there is some initial heat loss but temperature increases gradually, reaction rates will increase and may generate heat fast enough compared with heat loss to approach adibaticity. Thus, provision for ample cooling is essential to guard against thermal explosions. It follows that steam pressure to jackets of vessels or to steam coils should be at the

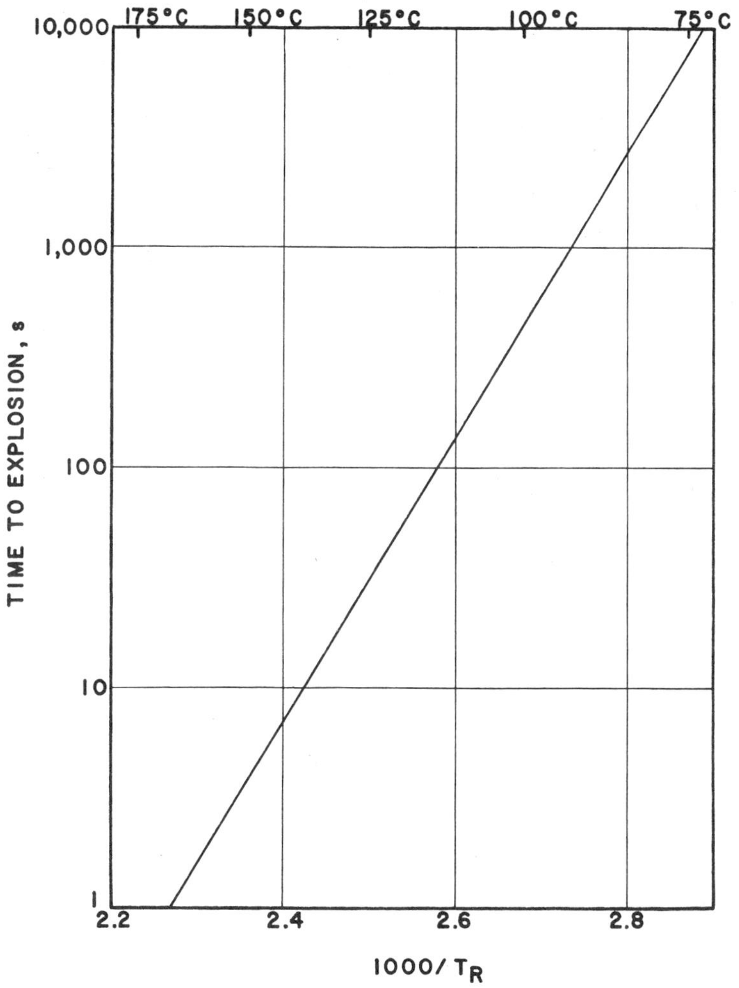

Fig. 7-1 Example of induction period for an adiabatic explosion for

$$
\begin{aligned}
c_p &= 1.5 \text{ kJ/kg·K} \\
\text{MW} &= 200 \\
Z &= 10^{13} \text{ s}^{-1} \\
\Delta H &= -1000 \text{ kJ/mol} \\
E_a &= 130 \text{ kJ/mol}
\end{aligned}
$$

lowest feasible pressure. Otherwise, a leak of high-pressure steam through a control valve might lead to a thermal explosion by producing an intolerably high jacket temperature. Boynton et al.[27] showed for stirred reactors that when one equates Eqs. (7-1) and (7-2) and differentiates with respect to T_R, the maximum stable temperature difference $(T_R - T_A)_{\text{crit}}$ equals $R_u T_R{}^2/E_a$. Once an exothermic reaction develops a temperature more than $R_u T_R{}^2/E_a$ above the surrounding temperature, a thermal explosion can occur. An example based on their work follows.

Example

Two moles (0.4 kg) of a decomposable compound is placed in a jacketed test apparatus. A thermal explosion occurs when the contents are held long enough at a jacket temperature of 150°C

$$T_R - T_A = \frac{R_u T_R^2}{E_a}$$

$T_R = 435$ K ($t_r = 162$°C) for $E_a = 130$ kJ/mol, and so $(t_r - t_a)_{crit} =$ 12°C. A thermal-stability diagram for this reaction is shown in Fig. 7-2. [$Z = 10^{13}$/s and $\Delta H_d = -1000$ kJ/mol. Thus, $U'A'' = 0.43$ kJ/(s)(K).] Temperature behavior for various temperatures of the ingredients and the jacket can be examined with the help of Fig. 7-2. With a jacket temperature of 145°C, the temperature of the ingredients in the vessel will rise only to 150°C; if the initial temperature of the compound is as high as 173°C, it will cool back to 150°C. The mixture at *A* is stable. A greater temperature difference between the vessel contents and surroundings than 12°C is therefore safe up to 173°C, when $t_a = 145$°C. The contents above 173°C at *C* are unstable; heat generation begins to exceed heat loss, and a thermal explosion can develop. With an initial t_r of 184°C at *D*, jacket temperature must be 119°C to prevent a runaway decomposition. At $t_a = 150$°C the maximum stable temperature of the compound is 162°C, *B* in Fig. 7-2. (The temperature of the vessel contents at *B* is sometimes called the *temperature of no return.*) If, now, the jacket temperature is raised to 155°C, no equilibrium can occur. Either the reactants are consumed, or an explosion occurs.

With twice the heat-transfer area, the maximum safe jacket temperature increases to about 159°C. This can be determined by locating the point *B'* (171°C) on the heat-generation curve in Fig. 7-2, where the heat loss is twice that at *B*. The value of $(t_r - t_a)_{crit}$ remains approximately 12°C, and so $t_{a,\ crit} = 159$°C.

Scale factors do not affect $(t_r - t_a)_{crit}$, which equals $R_u T_R{}^2/E_a$ for large or small equipment. They do influence the potential for a thermal explosion, however, as discussed below.

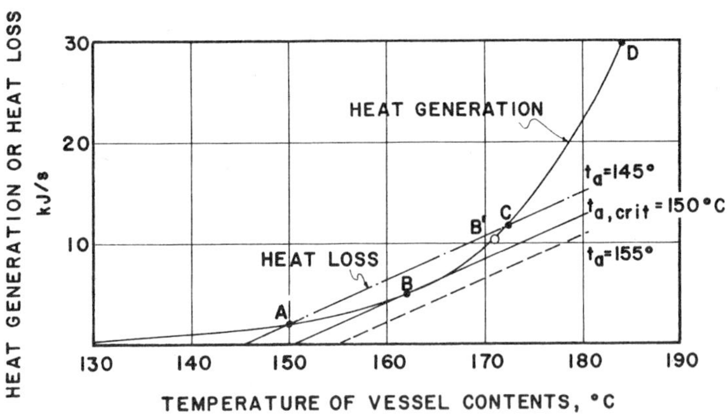

Fig. 7-2 Thermal-stability diagram, test apparatus.

For the larger equipment (subscript 2),

$$\frac{a_2}{a_1} = \frac{V_2}{V_1}$$

$$b = \frac{A_1''V_2}{A_2''V_1}$$

For 4000 mol (800 kg) of the decomposable material in a plant vessel, consider the case with $b = 12$. Heat generation increases by a factor of 2000 over the experimental equipment, and heat loss increases by 167. The reaction temperature for $(t_r - t_a)_{crit}$, determined from Eqs. (7-1) and (7-2), is

$$T_{R_2}{}^2 = T_{R_1}{}^2 b \exp\left(\frac{E_a}{R_u}\frac{T_{R_2} - T_{R_1}}{T_{R_2}T_{R_1}}\right) \tag{7-5}$$

Thus, $t_{r2} = 132°C$, and the maximum stable temperature difference is 10.5°C; the critical jacket temperature for the plant reactor is reduced to 121.5°C. A thermal-stability diagram for the plant equipment is shown in Fig. 7-3.

A larger reactor becomes unstable at a lower temperature than in the test apparatus. The rate of reaction at *B*, however, in Fig. 7-3 is only one-fifteenth the rate at *B* in Fig. 7-2. Nevertheless, with a jacket temperature of 130°C in Fig. 7-3, the temperature of the vessel contents is accelerating at the rate of about 0.5°C/s at $t_r = 135°C$ [specific heat = 0.3 kJ/(mol)(K)]. This can be determined from Fig. 7-3 using the net heat gain of 0.54 MJ/s. At 170°C, the temperature is increasing 13.5°C/s, and the system is close to adiabatic. At some higher temperature, the increase in the rate of decomposition will be balanced by the lessening in rate due to decrease in concentration. This is the maximum rate of decomposition.

More complex relationships and methods are necessary to handle decomposable solids quantitatively; they are discussed by Boynton et al.,[27] Stull,[33] and Townsend.[35]

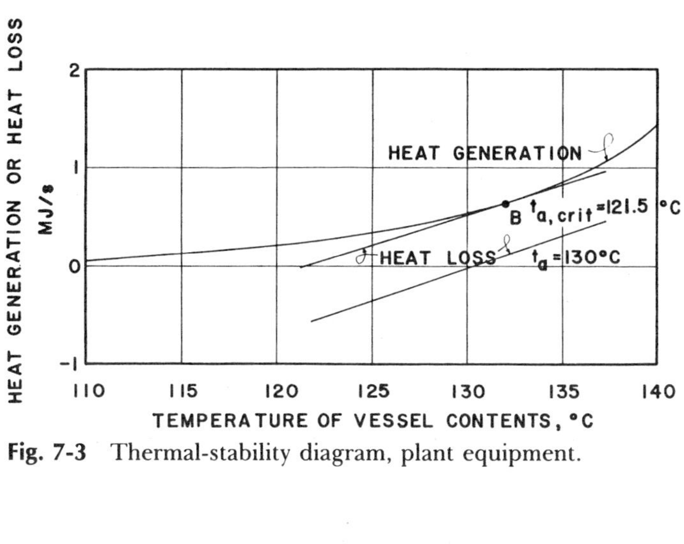

Fig. 7-3 Thermal-stability diagram, plant equipment.

Testing

A large variety of test procedures are available to ascertain the hazards of compounds and reactions. Generally, no single method can define the hazard of a system adequately. When tests are performed, the thermal environment of plant facilities should be duplicated to the extent feasible and the results of tests must be applied properly to the design and operation of plant equipment. Some tests on thermal stability plus mechanical and explosive shock are covered in the following sections. Indexes based on these tests and other parameters are also discussed.

Thermal stability. The ASTM has published three standard methods on thermal stability (Table 7-1). Others are under development.

E476 defines the temperature at which a reaction will start with generation of appreciable heat or pressure when subjected to a programmed temperature increase. The magnitude and rate of pressure generation can also be determined. It is to be used in conjunction with other tests. E487 determines the constant-temperature stability of chemical materials that undergo exothermic reactions without application of external heating. It does not determine a safe operating temperature. E537 uses techniques of differential thermal analysis (DTA) and differential scanning calorimetry (DSC) to detect enthalpic changes and to approximate the initiation temperatures of these changes. It is recommended by ASTM as an early test for detecting the reactive hazards of an uncharacterized chemical substance or mixture.

Thermally unstable substances that decompose exothermically do so at all temperatures, even though the rate of decomposition may be imperceptible at normal room temperature. It is often desirable to know the temperature at which perceptible self-heating starts; in an adiabatic system the heating may develop into a thermal explosion, but where heat loss is adequate, problems do not necessarily occur. The onset of an exotherm in DTA and DSC depends on the heating rate. Differential scanning calorimetry can be used, however, to determine E_a and Z. (A typical DSC trace is shown in Fig. 7-4, but the methodology of DSC is beyond the scope of this text.)* Then, a thermal-stability diagram can be constructed using Eqs. (7-1) and (7-2) and engineering estimates of

TABLE 7-1 ASTM Standard Test Methods on Thermal Stability

Title	Designation
Thermal Instability of Confined Condensed Phase Systems (Confinement Test)	ASTM E476-73
Constant-Temperature Stability of Chemical Materials	ASTM E487-74
Assessing the Thermal Stability of Chemicals by Methods of Differential Thermal Analysis	ANSI/ASTM E537-76

*Thermal methods of analysis are covered in Ref. 36.

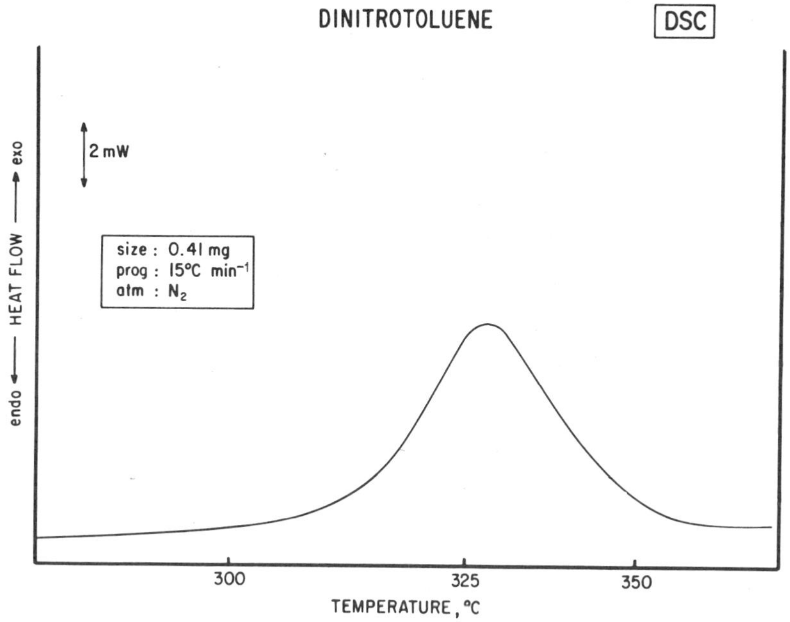

Fig. 7-4 DSC thermogram for the exothermic reaction of a mixed dinitrololuene isomer sample. Heat flow generated by the sample is plotted on the ordinate vs. linearly increasing sample temperature on the abscissa. [*E. I. du Pont de Nemours & Company, (Incorporated).*]

$U'A''$. The temperature corresponding to point B in Figs. 7-2 and 7-3 for agitated vessels can be found from

$$\frac{R_u T_R{}^2}{E_a} \exp \frac{E_a}{R_u T_R} = \frac{aZ\, \Delta H}{U'A''} \tag{7-6}$$

An actual exotherm may start below the temperature of the exothermic reaction recorded by DSC. Nevertheless, thermal analysis by DSC provides a powerful and expanding tool to help in evaluating thermal stability, thereby determining adiabatic behavior; for example, τ_{ad} can be calculated from Eq. (7-4). Also, half-life times, $\tau_{1/2}$ in seconds, can be calculated for first-order reactions as follows:

$$\tau_{1/2} = \frac{0.693}{k} = \frac{0.693}{Z} \exp \frac{E_a}{R_u T_R} \tag{7-7}$$

In addition, an accelerating rate calorimeter (ARC) has been developed to simplify hazard evaluation of self-accelerating reactions.[35] Comparatively crude tests to determine whether an exotherm occurs using a melting-point apparatus, a 25-cm³ stainless-steel bomb, and a Dewar flask are described in Refs. 37 to 39, respectively. The temperature at which decomposition would begin in plant equipment, however, is only approximated in these experiments; of the three tests, the Dewar flask probably best duplicates the adiabatic conditions that can occur in an insulated and jacketed plant reactor.

Mechanical and explosive shock. Tests on sensitivity to shock have been described by Stull[33] and Van Dolah.[40] The three principal ones are summarized in Table 7-2.

Commercial laboratories are available to perform shock tests. The ASTM is developing standard test methods on mechanical and explosive shock. Even if tests and additional analysis show detonation to be unlikely, sufficient energy might be released to develop high enough pressure to rupture a vessel.

Indexes of reactivity hazards. A summary of the NFPA ratings on the susceptibility of materials to release energy is given below.[41]

4 Materials which in themselves are readily capable of detonation or of explosive decomposition or reaction at normal temperatures and pressures.

3 Materials which in themselves are capable of detonation or explosive reaction but require a strong initiating source or which must be heated under confinement before initiation or which react explosively with water.

2 Materials which in themselves are normally unstable and readily undergo violent chemical change but do not detonate. Also materials which may react violently with water or which may form potentially explosive mixtures with water.

TABLE 7-2 Tests on Mechanical and Explosive Shock

Test	Description
Impact sensitivity	Sample subjected to impact of falling weight; Bruceton up-and-down technique generally used to estimate height of drop corresponding to 50% probability that ignition will occur[40]; consequences of ignition not determined; main use to compare sensitivity of various materials to ignition by impact; ASTM D2540, Standard Test Method for Drop-Weight Sensitivity of Liquid Monopropellants, covers determination of sensitivity of liquid monopropellants to impact of drop weight. ASTM E680 is for solids.
Trauzl	Used for compounds that detonate; 10-g sample placed in lead block is exploded; enlargement of cavity containing sample is reported as percentage of expansion caused by same weight of TNT
Card gap	Standard tetryl pellet used to create shock, which is attenuated by stack of cellulose acetate disks before passage into sample; severity of detonation judged by damage to steel witness plate, resting on top of sample container; results comparative; most widely used shock test; materials generally show greatest sensitivity with card-gap test; ASTM D2539 describes a Standard Test Method for Shock Sensitivity of Liquid Monopropellants by the Card-Gap Test

1 Materials which in themselves are normally stable, but which can become unstable at elevated temperatures and pressures or which may react with water with some release of energy but not violently.

0 Materials which in themselves are normally stable, even under fire exposure conditions, and which are not reactive with water.

Several hazard indexes have been developed for rating a particular compound or reaction according to the NFPA system. Some indexes are summarized in Table 7-3. Others are discussed by Treweek et al.[42]

It is difficult to select any one criterion from the array of indexes in Table 7-3 to appraise chemical instability. Nevertheless, if T_d, based on the maximum heat release, is more than 950 K, the possibility of detonation with NFPA ratings of 3 or 4 should be suspected; for peroxides the T_d value is 850 K or more for detonation potential. (T_d can be deter-

TABLE 7-3 Summary of Indexes of Reactivity Hazards

Description			Reference
Various + and − assignments given to thermal stability, impact sensitivity, and heat of reaction, as follows:			Coffee[43]
	+	−	
Thermal stability	DTA exotherm	No DTA exotherm	
Impact sensitivity, kg·m*			
Solids	<6.34	>6.34	
Liquids	<1	>1	
Heat of reaction, MJ/kg	<-2.93	>-2.93	
Rating of ++ or +++ for given compound or reaction indicates potential chemical instability; further evaluation should be performed, if only plus is $\Delta H < -2.93$ MJ/kg[43]			
Shock sensitivity predicted by CHETAH if $\Delta H_d \leq -2.93$ MJ/kg or either (1) $\Delta H_d < -1.26$ MJ/kg and numerical value of $\Delta H_c - \Delta H_d < 20.9$ MJ/kg or (2) oxygen balance† between −160 and 240			Seaton et al.[19] Treweek and Seaton[20]
Reaction-hazard index (RHI) $= \dfrac{10T_d}{T_d + 7.2E_a}$			Stull[45]
Adiabatic decomposition temperature T_d‡			DeHaven[46]
Logarithm system $\approx 1.3 + \log Z - 0.43 \dfrac{E_a}{R_u T_d}$ where Z = Arrhenius preexponential factor			DeHaven[46]

*Impacts of 6.34 kg·m (solids) or 1 kg·m (liquids) are more than occur practically in general.[43]

$$\text{†Oxygen balance}^{44} = \frac{-1600(2n + x/2 - y)}{\text{MW}}$$

where n = carbon atoms, x = hydrogen atoms, and y = oxygen atoms in $C_nH_xO_yN_z$.

‡Calculated temperature resulting from adiabatic decomposition starting at 298 K.

mined from an output of CHETAH plus other programs summarized by Treweek et al.[42])

The pressure developed in a closed vessel from decomposition is given by

$$P_m = \frac{P_i n_f T_d}{n_i T_i} \qquad (7\text{-}8)$$

With decomposition of liquids or solids there can be an extraordinary increase in the moles of gas. Coupled with high adiabatic decomposition temperatures, enormous pressures are possible. (Equivalent hydrostatic pressures resulting from detonation of condensed explosives are covered in Chap. 4.)

Finally, size and confinement of the test specimen, among other parameters, relative to plant scale make it difficult to make an ironclad assessment of safety when large amounts of energy can be suddenly released. Evaluation of the risks and consequences, considering potential ignition stimuli, by fault-tree analysis can improve that safety assessment, however.

7-3 Hazardous Operations

Several types of industrial operations pose fire and explosion risks which require evaluation. Some potentially hazardous operations are discussed in the balance of this section. The list is not intended to be all-embracing; other problems can lurk that require correction.

Compressors and Pumps

Fires and explosions have occurred in compressed-air systems, although these phenomena are not common. Generally, the incidents occur with oil-lubricated reciprocating compressors.[47] Carbonaceous deposits in the vicinity of the final-stage outlet valves and oil films appear to be the principal sources of combustibles in air-compressor installations.[48] Excessive carbon deposits may result from an improper oil feed rate; if oil deposits are heavy, the oil feed rate should be reduced. Also, intake air should be clean and cool to help reduce deposits on valves. An elevated intake normally is the best choice, with ducts leading to air filters at a lower level for accessibility.[47]

Ignition can occur spontaneously from adiabatic compression of air already in the system by a "piston" of the pressurizing air and by faulty final-stage outlet valves that produce sufficiently high temperature to ignite carbonaceous residues in the vicinity.[47,48] Ignition can also occur in normal operation if the final-stage outlet temperature is too high. Oil-film detonations can occur at pressures as low as about 700 kPa and possibly lower.[47] Burgoyne and Craven[47] have recommended the maximum safe outlet temperatures shown in Table 7-4.

TABLE 7-4 Maximum Recommended Safe Outlet Temperatures for Air Compressors[47]

Final-stage outlet pressure, MPa, gage	Maximum normal working temperature, °C	Automatic cutoff temperature, °C
1	145	155
2.5	137	147
5	130	140
10	124	134

In addition, periodic cleaning and preventive-maintenance schedules should be implemented to reduce carbonaceous deposits and to check the valves, valve seats, and other parts of the air-compressor system.

Fire-resistant phosphate ester oils have higher flash points and autoignition temperatures than mineral oils; they have been used successfully in compressed-air systems where the risk associated with use of mineral oils is not acceptable and where high final-stage outlet temperatures are unavoidable.[47,49] Problems can arise with use of phosphate ester oils because of their strong solvent action on paints and some types of rubber, which swell. Thus, as a general precaution, the compressor manufacturer should be consulted before phosphate ester oils are used.[47] Air compressed by fire-resistant fluid-lubricated compressors should not be used for air-line respirators or air-supplied suits.

When flammable gases are compressed, air should first be purged out of the system with inert gas. Otherwise, a flammable-gas–air explosion can occur; higher compression temperatures result with air than with flammable gases, which generally have relatively low heat-capacity ratios κ, as noted in Chap. 3. Also, special care is needed to avoid high temperatures when endothermic gases, such as acetylene, are compressed.

A potentially dangerous condition can occur if a centrifugal pump is run deadheaded. Under some circumstances ignition can occur when a flammable liquid is pumped by a conventional centrifugal pump against an air pocket in a closed line.[50] Adiabatic compression of the gas pocket may develop temperatures above the autoignition temperature of many fuels in air or possibly cause decomposition of a thermally unstable liquid. In addition, the liquid can be heated quickly in a deadheaded condition to a potentially dangerous temperature if thermally unstable liquids are being pumped. A thermally unstable material in a centrifugal pump can be protected against a hazardous temperature by installation of a temperature sensor in the pump as close to the impeller as possible. Alternatively, temperature sensors can be installed in the discharge piping from the pump, but in this case there must be provision to maintain flow past the sensor. In addition, use of a flow switch to shut down the pump if the flow drops to 10 to 20 percent of normal, for instance, should be considered.

Sight Glasses and Flexible Hoses

Rupture of sight glasses reportedly has been the cause of at least three serious explosions.[51–53] In one case vinyl chloride vapors and in another styrene vapors under pressure were released into a building.[51,52] Ignition sources were not clearly identified. These incidents highlight the need to question the advisability of sight glasses in similar pressure operations. When glasses are used, maintenance procedures should be implemented to assure proper installation. Moreover, the design must prevent strong thermal and hydraulic shocks to the sight glass. Also, a quick-closing and preferably automatically actuated valve ought to be installed between the vessel and the glass.[51]

Release of liquefied flammable gases from flexible hoses has caused explosions and fires; rubber hoses have burst, and leaks have occurred through couplings.[54] Special care is essential in transferring liquefied flammable gases. Couplings should be clean and unworn. Connections should be tight. Crushing and severing of the hoses must be guarded against. Wire braid used for reinforcement must be corrosion-resistant; a flexible stainless-steel line may be used as an alternative. Also, the hoses and couplings should be inspected regularly. Additional safety items on the transfer of liquefied flammable gases, including flexible connectors, are covered in Refs. 54 and 55.

Distillation Columns

During startup and shutdown of distillation columns, concentrations of flammable vapors may go through the flammable range. Also, in some emergency conditions, concentrations of vapors may become flammable; explosions have occurred in these circumstances, although they are rare. Even though the column can contain the explosion, internal damage to plates can be extensive. Thus, in some cases it may be prudent to purge columns with inert gas at startup and shutdown. (Where feasible, columns may be started with water instead of inert gas to prevent formation of flammable mixtures when coming on stream.) A pressure regulator on the column may be used to introduce nitrogen into the column at a few millimeters (+100 to 150) water column pressure to prevent subatmospheric pressure. A fluid seal set at slightly higher pressure will conserve inert gas. A pressure alarm should be provided on the inert-gas supply to warn of loss of inert gas. When vacuum distillation is performed, it is preferable to break vacuum with inert gas.

Distillation columns are often equipped with safety relief valves to protect the vessel, e.g., in event of a fire surrounding it. When processing thermally unstable materials, the pressure setting must be well below the equilibrium pressure corresponding to the unstable temperature.

Miscellaneous

Hydrogen is generated whenever batteries are charged. Loose terminals may cause sparking and an explosion during charging; checks should be made to assure tight connections. Nonconductive battery terminal covers can be used to prevent a short circuit of the battery and resultant sparks. They are generally available from the battery supplier or automotive parts dealers.

Heating of a relatively low-boiling liquid by an immiscible hot fluid can cause copious generation of vapor and entrained fluid, i.e., slop-over. (When the substances are mixed, their vapor pressures are additive.) Generally, this slop-over occurs when hot oil is added to water. If the vent is not large enough, the vessel may burst. In any event, the emission of oil causes a fire threat. To protect from incidents of this nature,[56]

> the incoming oil should be cooled *below* 100°C and a high-temperature alarm provided. If this is not possible because the oil will be too viscous, then the tank should be kept *above* 100°C so that any water present is vaporized. In addition, before the hot oil is added, the water layer should be drained off.

Explosions have occurred in plant and municipal sewers. Flying manhole covers are an obvious safety threat to personnel. Besides, they may break pipelines carrying flammable liquids and thus present a secondary fire and explosion hazard. Combustible-gas analyzers can be used to warn of the imminence of potentially hazardous concentrations of flammable gases in sewers. Additional safety items for consideration in the design of sewers and drains are covered in Ref. 50.

Sodium hydroxide is not usually considered a particularly hazardous chemical, but it has caused several explosions. For example, it causes violent condensation of aldehydes such as formaldehyde, acetaldehyde, and acrolein. It reacts with trichloroethylene to form explosive mixtures of dichloroacetylene; a pail of trichloroethylene dumped into a tank of caustic caused a fireball and eruption of the contents.[12] Furthermore, sodium hydroxide and other inorganic bases form explosive *aci*-nitro salts with nitroparaffins. It also forms explosive *aci*-nitro salts with aromatic compounds, such as *o*-nitrotoluene.[8] Thus, particular caution must be exercised to prevent contamination of processes and products by sodium hydroxide and other inorganic bases.

References

1. Witte, L. C., J. E. Cox, and J. E. Bouvier, "The Vapor Explosion," *J. Met.*, vol. 22, no. 2, pp. 39–44, February 1970.

2. Genco, J. M., and A. W. Lemmon, Jr., "Physical Explosions in Handling Molten Slags and Metals," *Trans. Am. Foundrymen's Soc.*, vol. 78, pp. 317–323, 1970.

3. Reid, R. C., "Superheated Liquids," *Am. Sci.,* vol. 64, pp. 146–156, March-April 1976.

4. Porteous, W. M., and R. C. Reid, "Light Hydrocarbon Vapor Explosions," *Chem. Eng. Prog.,* vol. 72, no. 5, pp. 83–89, May 1976.

5. Witte, L. C., and J. E. Cox, "The Vapor Explosion: A Second Look," *J. Met.,* vol. 30, no. 10, pp. 29–35, October 1978.

6. Nelson, W., "A New Theory to Explain Physical Explosions," *Tappi,* vol. 56, no. 3, pp. 121–125, March 1973.

7. Katz, D. L., "Superheat-Limit Explosions," *Chem. Eng. Prog.,* vol. 68, no. 5, pp. 68–69, May 1972.

8. Bretherick, L., *Handbook of Reactive Chemical Hazards,* 2d ed., CRC Press, Cleveland, 1979.

9. National Fire Protection Association, Hazardous Chemicals Data, *NFPA* 49, Boston, Mass.

10. National Fire Protection Association, Manual of Hazardous Chemical Reactions, *NFPA* 491M, Boston.

11. Manufacturing Chemists' Association, *Case Histories of Accidents in the Chemical Industry,* vol. 1, *Accident Case Histories 1–596,* Washington, 1962.

12. Manufacturing Chemists' Association. *Case Histories of Accidents in the Chemical Industry,* vol. 2, *Accident Case Histories 597–1097,* Washington, 1966.

13. Manufacturing Chemists' Association, *Case Histories of Accidents in the Chemical Industry,* vol. 3, *Accident Case Histories 1098–1623,* Washington, 1970.

14. Manufacturing Chemists' Association, *Case Histories of Accidents in the Chemical Industry,* vol. 4, *Accident Case Histories 1624–2108,* Washington, 1975.

15. Manufacturing Chemists' Association, *Guide for Safety in the Chemical Laboratory,* 2d ed. Van Nostrand Reinhold, New York, 1972.

16. Mellor, J. W., *A Comprehensive Treatise on Inorganic and Theoretical Chemistry,* vols. I–XVI, Longmans, Green, London, 1922 1937.

17. Kirk, R. E., and D. F. Othmer, *Encyclopedia of Chemical Technology,* 2d ed., vols. 1–22, suppl. vol., and index vol., Interscience, New York, 1963–1972.

18. Stull, D. R., and H. Prophet, JANAF Thermochemical Tables, *Natl. Bur. Stand. (U.S.)* NSRDS-NBS37, Washington, 1971.

19. Seaton, W. H., E. Freedman, and D. N. Treweek, CHETAH: The ASTM Chemical Thermodynamic and Energy Release Evaluation Program, *ASTM Data Ser. Publ.* DS51, November 1974.

20. Treweek, D. N., and W. H. Seaton, "Appraising Energy Hazard Potentials," *Chem. Eng. Prog. 7th Loss Prev. Symp., New York, 1973,* pp. 21–27.

21. Deason, W. R., W. E. Koerner, and R. H. Munch, "Determining Maximum Heat Load in Equipment Design; Determination of Heat Evolution Rates; Effects of Ring Substituents and Added Contaminants on Nitrobenzenes," *Ind. Eng. Chem.,* vol. 51, no. 9, pp. 1001–1004, September 1959.

22. Armitage, J. B., and H. W. Strauss, "Safety Considerations in Industrial Use of Organic Peroxides," *Ind. Eng. Chem.,* vol. 56, no. 12, pp. 28–32, December 1964.

23. Santini, D. G., "Factors Influencing the Design and Operation of a Peracetic Acid Unit," *Chem. Eng. Prog.,* vol. 57, no. 12, pp. 61–65, December 1961.

24. Jackson, H. L., W. B. McCormack, C. S. Rondestvedt, K. C. Smeltz, and I. E. Viele, "Control of Peroxidizable Compounds," *J. Chem. Educ.,* vol. 47, no. 3, pp. A175–A188, March 1970.

25. Daniel, R. L., "Chlorination Hazards," *Chem. Eng. Prog. 7th Loss Prev. Symp., New York, 1973,* pp. 67–77.

26. Statesir, W. A., "Explosive Reactivity of Organics and Chlorine," *Chem. Eng. Prog. 7th Loss Prev. Symp., New York, 1973,* pp. 114–120.

27. Boynton, D. E., W. B. Nichols, and H. M. Spurlin, "How to Tame Dangerous Chemical Reactions," *Ind. Eng. Chem.,* vol. 51, no. 4, pp. 489–494, April 1959.

28. Monger, J. M., H. J. Baumgartner, G. C. Hood, and C. E. Sanborn, "Detonations in Hydrogen Peroxide Vapor," *J. Chem. Eng. Data,* vol. 9, no. 1, pp. 124–127, January 1964.

29. Kuchta, J. M., G. H. Martindill, M. G. Zabetakis, and G. H. Damon, "Flammability and Detonability Studies of Hydrogen Peroxide Systems Containing Organic Substances," *U.S. Bur. Mines Rep. Invest.* 5877, 1961.

30. Concentrated Hydrogen Peroxide, Summary of Research Data on Safety Limitations, *Shell Chemical Corp. Bull.* SC59–44, 1959.

31. Monger, J. M., H. Sello, and D. C. Lehwalder, "Explosion Limits of Liquid Systems Containing Hydrogen Peroxide, Water, and Oxygenated Organic Compounds," *J. Chem. Eng. Data,* vol. 6, no. 1, pp. 23–27, January 1961.

32. Pickles, R. G., "Hazard Reduction in the Formaldehyde Process," *Inst. Chem. Eng. Symp. Ser.* 34, pp. 57–60, 1971.

33. Stull, D. R., Fundamentals of Fire and Explosion, *AICHE Monogr. Ser.,* vol. 73, no. 10, 1977.

34. Frank-Kamenetskii, D. A., *Diffusion and Heat Transfer in Chemical Kinetics,* 2d ed., J. P. Appleton (trans. ed.), Plenum, New York, 1969.

35. Townsend, D. I., "Hazard Evaluation of Self-Accelerating Reactions," *Chem. Eng. Prog.,* vol. 73, no. 9, pp. 80–81, September 1977.

36. Wendlandt, W. W., *Thermal Methods of Analysis,* 2d ed., vol. 19 of *Chemical Analyses: A Series of Monographs on Analytical Chemistry and Its Application,* Wiley, New York, 1974.

37. Albisser, R. H., and L. H. Silver, "Safety Evaluation of New Processes," *Ind. Eng. Chem.,* vol. 52, no. 11, pp. 77A–79A, November 1960.

38. Rapean, J. C., D. L. Pearson, and H. Sello, "A Test for Hazardous Chemical Decomposition," *Ind. Eng. Chem.,* vol. 51, no. 2, pp. 77A–78A, February 1959.

39. Fawcett, H. H., and W. S. Wood, *Safety and Accident Prevention in Chemical Operations,* Interscience-Wiley, New York, 1965.

40. Van Dolah, R. W., "Evaluating the Explosive Character of Chemicals," *Ind. Eng. Chem.,* vol. 53, no. 7, pp. 59A–62A, July 1961.

41. National Fire Protection Association, Standard System for the Identification of the Fire Hazards of Materials, *NFPA* 704, Boston.

42. Treweek, D. N., J. R. Hoyland, C. A. Alexander, and W. M. Pardue, "Use

of Simple Thermodynamic and Structural Parameters to Predict Self-Reactivity Hazard Ratings of Chemicals," *J. Hazard. Mater.,* vol. 1, no. 3, pp. 173–189, November 1976.

43. Coffee, R. D., "Evaluation of Chemical Stability," *J. Chem. Educ.,* vol. 49, no. 6, pp. A343–A349, June 1972.

44. Lothrop, W. C., and R. Hendrick, "The Relationship between Performance and Constitution of Pure Organic Explosive Compounds," *Chem. Rev.,* vol. 44, no. 3, pp. 419–445, June 1949.

45. Stull, D. R., "Linking Thermodynamics and Kinetics to Predict Real Chemical Hazards," *Chem. Eng. Prog. 7th Loss Prev. Symp., New York, 1973,* pp. 67–73.

46. DeHaven, E. S., "Using Kinetics to Evaluate Reactive Hazards," *Chem. Eng. Prog. 12th Loss Prev. Symp., Atlanta, 1979,* pp. 41–44.

47. Burgoyne, J. H., and A. D. Craven, "Fire and Explosion Hazards in Compressed Air Systems," *Chem. Eng. Prog. 7th Loss Prev. Symp., New York, 1973,* pp. 79–87.

48. Perlee, H. E., and M. G. Zabetakis, "Compressor and Related Explosions," *U.S. Bur. Mines Rep. Invest.* 8187, 1963.

49. Anon., "Maintenance Notes," *Maintenance,* December 1959, pp. 11–12.

50. *Inst. Chem. Eng. Loss Prevention Bull.* 019, London, 1978.

51. Walls, W. L., "Vinyl Chloride Explosion," *Natl. Fire Prot. Assoc. Q.,* vol. 57, no. 4, pp. 352–362, April 1964.

52. Anon.: "Fatal Chemical Plant Explosion," *Fire J.,* vol. 61, no. 5, pp. 8–11, September 1967.

53. Adcock, C. T., and J. D. Waldon, "To Minimize the Loss in a Chemical Plant Explosion," *Saf. Maint.,* vol. 134, pp. 45–48, October 1967.

54. *Inst. Chem. Eng. Loss Prev. Bull.* 013, London, 1977.

55. National Fire Protection Association, Standard for the Storage and Handling of Liquefied Petroleum Gases, *NFPA* 58, Boston.

56. Kletz, T. A., "Accidents That Will Occur during the Coming Year," *Chem. Eng. Prog. 10th Loss Prev. Symp., Kansas City, 1976,* pp. 151–154.

Additional References

Reid, R. C., "Possible Mechanism for Pressurized-Liquid Tank Explosions or BLEVE's," *Science,* vol. 23, pp. 1263–1265, Mar. 23, 1979.

Henry, R. E., and H. K. Fauske, "Nucleation Processes in Large Scale Vapor Explosions," *Trans ASME J. Heat Trans.* vol. 101, no. 2, pp. 280–287, May 1979.

8

Dust Explosions

As manufacturing plants have become larger, dusts and powders are being handled in increasing quantities. Complex pneumatic handling systems with large silos pose relatively new and special fire and explosion risks.[1,2] Also, the advent of powder coating requires proper design and operation to guard against explosion and fire. Thus, knowledge of the fundamentals on the causes of dust explosions and of the procedures for prevention and protection is essential for safe operation of plants where dusts are processed.

8-1 Explosible Concentrations

Explosibility Limits

Combustible solids may explode if they are mixed in the right concentration in air. Pressure in a closed space results from heat released during rapid combustion; the change in moles of gases resulting from combustion ordinarily is small, and consequently such change does not contribute significantly to increase of pressure. Generally, the minimum explosible concentration (MEC) of fine solid organic materials in air is about 20 mg/L. This resembles a very dense fog in appearance, and it takes a fairly large quantity of dust to give that concentration. (Consider a 30-m^3 space. It would take 2 L of powder with a bulk density of 300 kg/m^3 to give a powder concentration of 20 mg/L in 30 m^3 of space.) The Bureau of Mines has investigated the explosibility of a host of dusts. Data are given in Refs. 3 to 6, for example. The criterion for the MEC is the development of sufficient pressure, that is, 14 to 21 kPa, to burst a filter-paper diaphragm in the test apparatus.[7]

The Bureau of Mines data are for dust through a 200-mesh screen (74 μm diameter). The effect on the MEC is pronounced for dust coarser than through a 200-mesh sieve but is negligible for finer dusts.[3] Thus, the MEC of fine cellulose acetate (minus 200-mesh) is approximately 30 mg/L, but it increases to about 200 mg/L for an average particle diameter of 300 μm.[8] In general, particles larger than about 500 μm (dust on a 35-mesh screen) do not explode.[6] Attrition, however, can produce fine dust from coarse dust.

Carbonaceous dusts, such as activated carbon, asphalt, and coal, with low volatile content are not explosion threats; in general, carbonaceous dusts with a volatile content of 13 percent or less cannot be ignited by a continuous induction spark in which 24 to 28 W is dissipated.[5,7]

The upper explosibility limits (UEL) of dusts are poorly defined, and not many tests on UELs have been performed because of the difficulties in doing the tests with such high concentrations and the limited applications of the data. Flame was quenched by a coal-dust concentration of 5 kg/m^3 in an experimental coal mine.[9] Thus, UELs, in general, are about 4 kg/m^3 (4×10^3 mg/L). Some materials may exhibit lower values, while a few others might have higher UELs.

Determination of Dust Concentrations

Dust mass balances and fan capacity can be used to estimate concentrations of dust in pipelines. Thus, 100 kg/h of a dust in 1 m^3/s air results in a concentration of 27.8 mg/L, which could be explosible. An additional 1 m^3/s air will lower the concentration to 13.9 mg/L, which is more than likely not explosible.

When dust flows into a silo, for instance, a collector is used first to collect the dust, which then falls into the silo. Air leaves through the collector. The resulting average concentration in the silo can be approximated by

$$\chi_D = \frac{0.28Q_D}{A'v_t} \tag{8-1}$$

Terminal velocity v_t, can be determined from Ref. 10.

Example

A dust with a specific gravity of 1.5 (H_2O = 1) and 74 μm in diameter (through 200 mesh) has a terminal velocity of 0.20 m/s. If it flows into a 6-m-diameter silo at the rate of 400 kg/h, the resulting average concentration is

$$\chi_D = \frac{(0.28)(400)}{(\pi/4)(6)^2(0.20)} = 19.8 \text{ mg/L}$$

Particle size has a disproportionately large effect on χ_D in Eq. (8-1). Thus, for particle diameters of 43 μm (through 325 mesh) and 246 μm (through 60 mesh), the terminal velocities for the above dust are 0.08 and 1.2 m/s, respectively. Corresponding χ_D's from Eq. (8-1) for the above conditions are 49.5 and 3.30 mg/L.

Moodie[11] has described dust-sampling procedures that have been used to aid in hazardous-dust-area classification. In addition, air quality-control sampling methods have been used to determine concentrations of dust within plant equipment, including silos and pipelines. A schematic diagram of dust-sampling apparatus for determining the dust concentration in a pipeline is shown in Fig. 8-1. Air velocity in the duct must match the speed through the tip of the sampling nozzle (isokinetic sampling). Otherwise sampling results will not be representative of concentrations in the pipeline. Additional guidance on sampling techniques is included in Refs. 12 and 13.

An optical dust-concentration probe is described in Ref. 14.

8-2 Ignition

Particle Size

Dusts have a much higher electrical minimum ignition energy (MIE) than flammable vapors or gases, as shown in Fig. 3-5. The MIE of dusts and powders is around 100 times that of flammable vapors.* (Dust suspensions wet with solvent or in a solvent-air atmosphere can be much easier to ignite than dry dust. For example, plastic liners in bags can result in sparks with sufficient energy to ignite flammable vapors when powder is emptied into a vessel containing a flammable liquid.) The MIE of a dust cloud is the least spark energy from a condenser required to produce flame propagation 100 mm or longer in a 305-mm-long test apparatus (volume = 1.23 L). MIE values are only approximations. Particle size has a profound effect on MIE; fine dust is much easier to ignite than coarse dust. Also, it is dispersed more easily and remains in suspension longer than coarse dust. The effect of particle diameter on MIE is

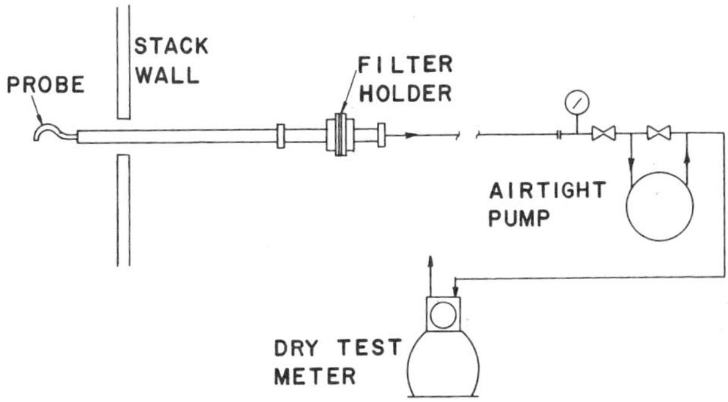

Fig. 8-1 Schematic diagram of a dust-sampling apparatus. (Equipment must be electrically grounded.)

*Sparks of long duration may ignite dusts with less energy than short, high-intensity sparks,[15–17] which apparently push the dust particles away from the spark region.[15,16] More information is needed on this phenomenon, however; see Ref. 31.

shown in Fig. 8-2. Plant ignition sources of sufficient energy to ignite dust with an MIE > 500 mJ are uncommon. Caution, nevertheless, must be exercised because classification by gravity can produce an ignitable cloud from the fine dust in a mixture possessing a relatively high overall average particle size and MIE. MIEs are determined at the most easily ignitable concentration, generally 5 to 10 times the MEC.[7] Consequently, ignition energies for marginally explosible concentrations are much higher than published MIEs.

Static Electricity

Electric charges are generated on dust particles when dust clouds are generated and when they move through plant equipment. Thus, dust can create its own ignition source. In almost all cases of dust explosions started by static electricity, sparking occurs between an insulated electrical conductor charged by the dust and nearby grounded equipment. *All dust-handling equipment should be grounded, either directly or by bonding to grounded equipment. Also, good housekeeping is imperative to minimize the chance of formation of a dust cloud and to limit fuel supply.* A spark from a conductor dissipates all the stored energy in that single discharge, while sparks from a charged insulator contain only a portion of the stored energy; accordingly they are relatively weak but can be relatively long in duration. Such sparks can ignite flammable vapors and gases and may ignite dusts under severe transfer conditions; Hughes and Bright[18] have reported occur-

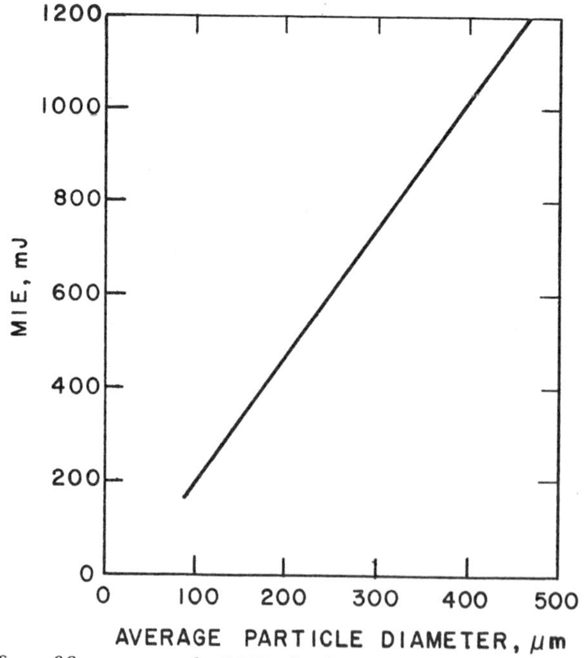

Fig. 8-2 Effect of fineness on the MIE of cellulose acetate molding powder. (*After Hartmann and Nagy.*[8])

rence of incendive sparks between polyvinyl chloride (PVC) pipe and a grounded metal collar around the pipe. These sparks occurred during the bottom loading of chocolate crumb through the PVC pipe into the silo. Since potentially dangerous electric charging occurs when transporting highly resistive dusts in plastic pipe, piping made from electrically insulating material should not be used to convey ignitable dusts or dusts that evolve flammable vapors (see Sec. 8-3).

Although strong electric charges often occur in dust storms in nature, it is unlikely that self-generated sparks can occur in a dust cloud unless it is at least the size of a small house.[19] Nevertheless, the density of electric charge from a chute can be relatively high. In this case there may be an electric discharge from the charged powder, even though it is a nonconductor, to a grounded pointed object, as shown in Fig. 8-3. Plant explosions have occurred from this effect. Thus, protrusions into the top of dust receivers should be avoided.

As indicated in Chap. 3, low humidity promotes static electrification because of the resulting decrease in moisture and in conductivity of nonconductors. This effect results in a higher frequency of fires and dust explosions in winter than in summer.[19,20] The relative observed frequency of occurrence of major and minor coal-dust explosions is shown in Fig. 8-4.

Temperature

High temperature can cause ignition of dust clouds and layers. The ignition temperature of a dust cloud is determined in a Godbert-Greenwald furnace; ignition of a dust cloud is denoted by the appearance of flame below the mouth of the furnace. Ignition-temperature data of the Bureau of Mines are for dust through a 200-mesh screen.[3,7] (The ignition temperature of a dust cloud is not profoundly affected by particle size.[3])

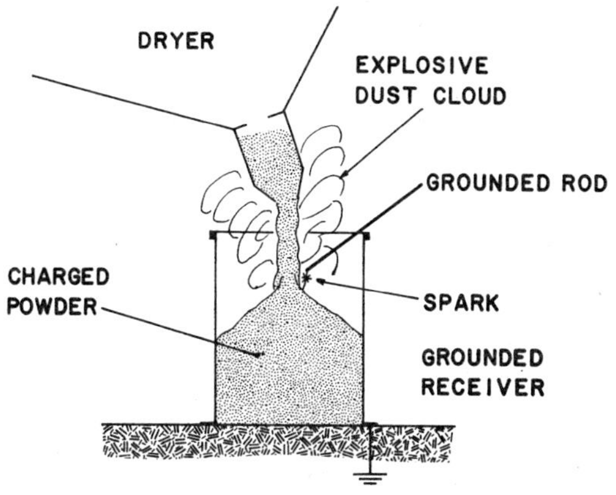

Fig. 8-3 Dangerous spark from surface of charged powder inside grounded container. *(After Cooper,[19] by permission. Copyright © by The Institute of Physics.)*

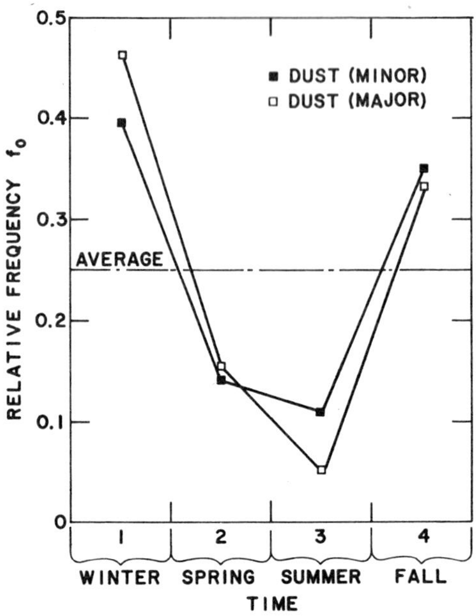

Fig. 8-4 Relative observed frequency of occurrence of major and minor dust explosions in coal mines. *(After Kissell et al.,*[20] *by permission. Copyright © 1973 by the American Association for the Advancement of Science.)*

The ignition temperature of a dust cloud is the hot-surface temperature in the furnace, not the lower (and unknown) temperature of the dust-air mixture in the furnace. Thus, it is not the autoignition temperature but is similar to the hot-surface ignition temperature for vapors and gases discussed under ignition delay and flow conditions in Sec. 3-1.

A modification of the Godbert-Greenwald furnace is used to determine the ignition temperature of a dust layer.[7] A sample of dust filling a 12.7-mm-deep, 25.4-mm-diameter container is used in the layer tests; ignition is denoted by a glow or flame. The ignition temperatures of a few plastic and agricultural dusts are shown in Table 8-1.

Natural materials tend to have lower ignition temperatures than manufactured chemicals. Also, dust layers generally exhibit lower ignition temperature than clouds. Moreover, the size of the sample affects the ignition temperature of a dust layer; ignition temperature decreases as the thickness of the layer increases, as shown in Fig. 8-5 for coal dust. (These tests were carried out on a hot plate with the top of the layer exposed to normal room temperature, in contrast to the much higher air temperature in the furnace layer tests just described; the minimum temperature of the plate at which ignition occurred is the ignition temperature. The criterion for ignition was smouldering combustion on the upper surface of the dust layer.) Palmer and Tonkin[21] have indicated that the minimum ignition temperature of this coal dust can be determined by extrapolation to $1/\lambda = 0$. Thus, the minimum ignition temperature for a deep layer of

TABLE 8-1 Dust Ignition Temperatures[3,4]

	Cloud, °C	Layer, °C
Plastic:		
Acrylonitrile polymer	500	460
Cellulose acetate from bag filter	460	430
Nylon polymer from filter	500	430
Polyethylene, high-pressure process	450	380
Polystyrene beads	500	470
Agricultural:		
Corn	400	250
Peanut hulls	460	210
Powdered sugar	370	400
Skim milk	490	200
Wheat flour	440	440

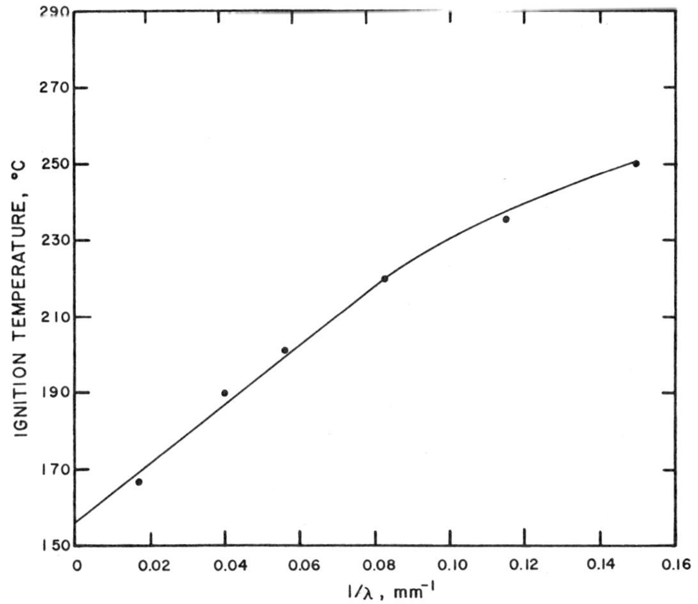

Fig. 8-5 Relation between the ignition temperature and depth of dust λ. *(After Palmer and Tonkin,[21] by permission.)*

coal dust, for instance, is 155°C, as shown in Fig. 8-5. The ignition temperature of a layer of coal dust is related to the logarithm of the depth of the layer between 2.90 and 59.0 mm, as shown in Fig. 8-6.

The relationship in Fig. 8-6 can be expressed as

$$t = 326.3 - 95.50 \log \lambda \text{ for } \lambda = 2.90 \text{ to } 59.0 \text{ mm} \qquad (8\text{-}2)$$

where t is the ignition temperature of the coal-dust laer in degrees Celsius.

Furthermore, the ignition temperatures of dust layers or of dust coat-

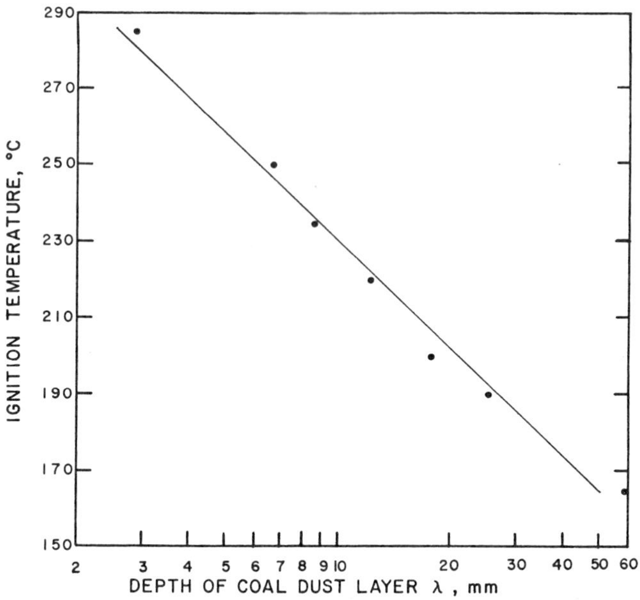

Fig. 8-6 Relation between the ignition temperature and logarithm of the depth of dust λ. (*Data of Palmer and Tonkin.*[21])

ings depend upon their thermal environment. *Ignition may occur when the rate of heat generation by autooxidation exceeds the rate of heat loss.* In dryers, for example, airflow should always be turned on when heat is on to promote heat loss from the dust. The larger the volume of the material the lower the temperature required to initiate perceptible self-heating and self-ignition.[22] Also, piles of solid material, as in storage vessels, can catch on fire if heat can be generated by oxidation where it is stored at too high a temperature. A fire may occur abruptly, too, when buried, smouldering material is exposed to air.

The depth of a hazardous layer of grain dust in which propagation of fire occurs from ignition by a heated wire is about 1.25 mm.[11]

Close attention must be paid to electric light bulbs. Ignition of dust by electric light bulbs has often been reported. Bulbs overheat when insulated with dust, but the hot glass can ignite the dust before the filament burns out.[23] Moreover, a loose bulb in a socket presents a double hazard, as it may combine arcing with the production of heat.[24]

8-3 Inert Gas

Except for metal powders, flame propagation of substantially all combustible dust clouds from electric sparks,* whatever the concentration of

* $\leq$ 24 to 28 W in a continuous induction spark.

dust, can be prevented by reduction of oxygen to 11% v/v using CO_2 as the inert gas and to 8% using N_2.[6,25] Metal powders require less oxygen for effective inerting. The limiting (and thus safe) percentage of oxygen, however, is higher for many dusts. Margins of safety are necessary in the operation of industrial plants. Limiting oxygen percentages to prevent ignition of dust clouds by a surface temperature of 850°C are around 5% v/v to O_2; for dusts containing carbon, 3 to 4% v/v to O_2 may be necessary for effective inerting against ignition by this high temperature in rare cases.[25]

MIEs increase as oxygen decreases. Thus, the MIE of a cornstarch dust cloud at a concentration of 1620 mg/L increases from about 20 mJ in air to approximately 500 mJ with 15% O_2 in an air–CO_2 atmosphere.[25] Therefore, inerting down to the limiting oxygen concentration of about 11% v/v for a high-voltage 24-W induction spark for this dust may not be essential if the chance of an ignition source of this strength is negligible; only partial inerting by the carbon dioxide and/or water vapor generated in a dust dryer, for example, may yield a mixture with such a high MIE that ignition is virtually impossible. Also, the minimum ignition temperature of a dust cloud increases as oxygen decreases.

There is a growing trend toward using inert gas in silos, bins, and hoppers; conveying air generally does not pass through a silo, and consequently the amount of inert gas for effective inerting is not impracticable. If they are overheated, some polymers and other materials may decompose and evolve flammable gases. Also, flammable vapors may be generated from dust that contains residual solvent. If flammable gases are or can be present, ample inerting will also prevent a gas explosion. Oxygen concentration should be monitored continuously. If oxygen reaches unsafe levels, operations producing dusting should be shut down. Thus, if inert gas is used in silos, dust conveying should be halted automatically at the preset oxygen value. Alternative monitoring procedures are discussed in Sec. 2-5.

8-4 Explosion Pressure

Maximum and near maximum explosion pressure and pressure rise for dusts occur within a concentration range of about 200 to 1000 mg/L. Thus, there is a broad range of dust concentrations exhibiting the severest explosion-pressure effects. In sharp contrast to vapors and gases, however, these peak pressure effects occur at concentrations much above the stoichiometric concentration in air. Thus, for benzoic acid (C_6H_5COOH), for example, the C_{st} is 139 mg/L air (25°C), and the peak explosion pressure effects for fine dust through a 200-mesh screen occur at a much higher concentration,[6] as shown in Fig. 8-7.

The peak absolute explosion pressure for most fine dust clouds is about 8 times the initial absolute pressure, i.e., the same for vapors and gases discussed in Chap. 4. The maximum explosion pressure is not substantially affected by particle size, but the maximum rate of explosion-pressure rise increases greatly as particle size decreases, as shown in Table 8-2 for cornstarch.

Thus, *fine dusts exhibit greater explosion-pressure effects than coarse dusts besides being easier to disperse and to ignite.*

It has not been established whether or not combustible dust-air mixtures can detonate in an industrial plant.[27] Nevertheless, normal combustible dust-air mixtures have not been known to detonate.[28] Coal-dust clouds in coal mines, however, have detonated where there was a considerable degree of *confinement* and lengthy shafts.[28]

Various inert diluents of the dust or air decrease explosion pressure and pressure rise.

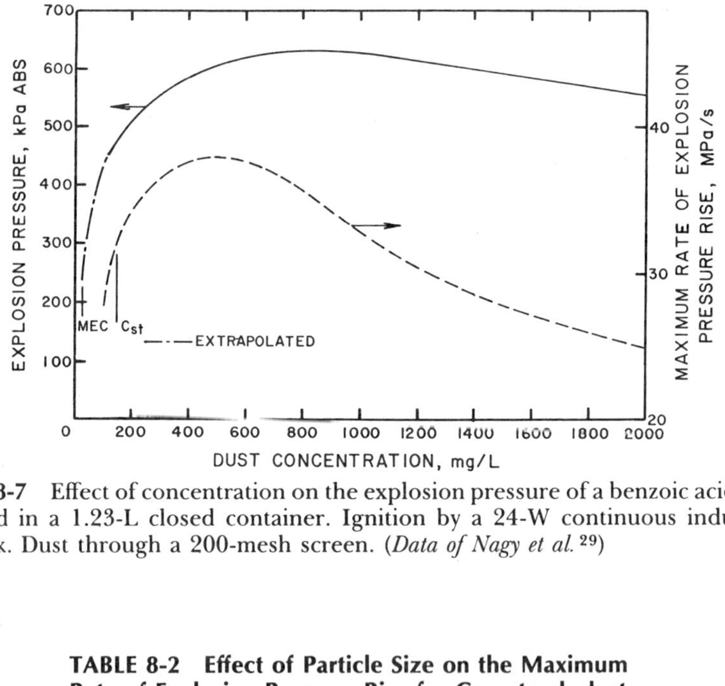

Fig. 8-7 Effect of concentration on the explosion pressure of a benzoic acid dust cloud in a 1.23-L closed container. Ignition by a 24-W continuous induction spark. Dust through a 200-mesh screen. (*Data of Nagy et al.* [29])

TABLE 8-2 Effect of Particle Size on the Maximum Rate of Explosion-Pressure Rise for Cornstarch-dust Cloud[26]

Average particle diameter, μm	Maximum rate of explosion-pressure rise, MPa/s*
178	1.72
126	6.20
89	13.1
37	45.8
22	59.3

*Cornstarch at 500 mg/L. Explosion test-vessel volume (unvented) = 1.23 L.

Moisture

For explosibility tests at the Bureau of Mines, a dust having more than 5 percent moisture is dried.[7] With spark ignition the percentages of water in dust required to prevent ignition of dust clouds varied from 16 percent for coal with a volatile content of 37 percent to 19 percent for cornstarch and 35 percent for paper dust.[25] Nevertheless, explosion severity is ameliorated with intermediate moisture contents, as shown in Fig. 8-8 for cornstarch. High moisture content may also inhibit creation of a dust cloud.

Inert Dust

The relative flammabilities of a dust cloud are defined as the percentage of inert dust (fuller's earth) required to suppress flame propagation by high temperature and by an electric spark.[7] Propagation is defined as the appearance of flame at the bottom of the test furnace with the heated surface at 700°C and at the bottom of a continuous-induction-spark tube (24 W). Stronger ignition sources are used in the tube if the dust cannot be ignited by the spark.[6] Very large amounts of inert dust are required to suppress flame propagation. As an example, the percentages of fuller's earth required to suppress flame propagation of cornstarch from ignition by the heated surface and the spark are more than 90, and 90 percent, respectively.[25] Even so, explosion pressure effects are substantially re-

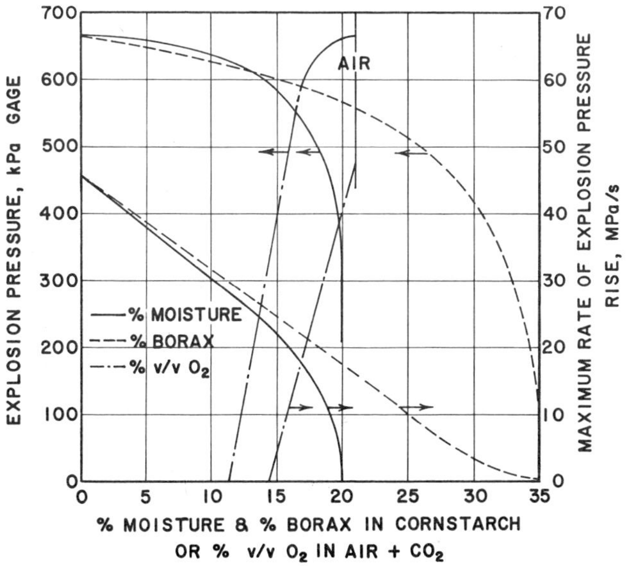

Fig. 8-8 Effect of moisture and inerts on explosion pressure and maximum rate of explosion-pressure rise of a cornstarch dust cloud at a concentration of 500 mg/L in a 1.23-L closed container. Ignition by a 24-W continuous induction spark. Dust through a 200-mesh screen. (*Data of Nagy et al.*[29])

duced by lesser amounts of inert dust, as shown in Fig. 8-8 for cornstarch with borax as the inert.[29] Small amounts of inert dust and moisture have only a small effect on maximum explosion pressure and maximum rate of explosion-pressure rise. As is evident from Fig. 8-8, moisture has a greater effect than inert dust on decreasing explosion intensity.

Oxygen

Sufficient decrease in oxygen can prevent ignition of dust clouds. Just partial inerting by inert gas increases the MIE, as noted in Sec. 8-3. Also, partial inerting reduces the maximum explosion pressure and maximum rate of explosion pressure rise, as shown in Fig. 8-8. The maximum rate of explosion-pressure rise decreases about linearly with decrease in the concentration of oxygen.[29]

8-5 Explosion Protection

One of the explosion-protection methods discussed in Chap. 5, i.e., containment, suppression, or venting, may be used for dusts as the sole method of reducing a dust-explosion hazard. Explosion-protection procedures can also be used as backup for explosion-prevention measures.

Secondary Explosions

With dusts and powders, safeguards against secondary explosions are vital. Secondary explosions can occur when a primary explosion in equipment dislodges settled dust in the surrounding rooms, which then explodes. Walls of these rooms usually cannot withstand much pressure and thus can fail at low pressure. Hence, the merit of good housekeeping. Moreover, damage to a building and injury to personnel can occur if plant equipment is explosion-vented into a room. Where feasible, dust handling and processing equipment should be installed outdoors. If it is installed indoors and explosion venting is practiced, the explosion products should be vented to the outdoors through as short a duct as possible. Thus, in planning new installations, equipment to be explosion-vented should be placed near outside walls if indoor installation is necessary.

Dust Handling

Certain unique procedures and equipment have to be used for handling dust because it is a solid material. Safe handling is discussed in the following paragraphs.

Sometimes scoops are used to transfer material from a drum to a vessel manually. A plastic or a grounded metal scoop should be used to prevent formation of a dangerous spark from static electricity. The grounded metal scoop has the disadvantage that the ground wire may become broken or disconnected. Consideration should also be given to use of Legstats* or Wristats* by the person doing the transfer. If it is eventually mixed with water anyway, powder can be wetted before transfer to reduce the risk of ignition.

Vacuum cleaners are sometimes used to clean up spills and for general housekeeping. Although they are rare, explosions in vacuum cleaners have occurred in industrial plants. All metal, including wands and wire in flexible hoses, should be grounded to minimize the chance of ignition. Requirements of the National Electrical Code must also be met. Explosion-protection techniques outlined in Chap. 5 often are not feasible with vacuum cleaners. Thus, large accumulations of powder should be prevented to the extent practical. If accumulations do occur, they can first be shoveled up or washed down to reduce the explosion potential in subsequent use of vacuum cleaners.

To reduce dust deposits in pipelines and ducts, conveying velocity should not be below about 20 m/s. Pipelines and ducts can usually be designed easily to contain the pressure from a dust explosion. Couplings sometimes turn out to be the weak link in implementing containment. Accordingly, the pressure rating of couplings should be checked to determine whether they meet design pressure requirements. On the other hand, pipelines or couplings can be made to fail intentionally at a particular location as a supplementary explosion vent as long as no damage or injury will result.

Explosions have occurred because of a metal object entering a grinding mill or pulverizer. Magnetic separators are available commercially and can be used to remove tramp metal. As a substitute, tramp-metal detectors can be used to warn of the presence of metal and to shut down processing equipment. Moreover, vibrating screens used to separate large dust particles will also remove large pieces of metal. In addition, construction, operating, and maintenance procedures should be adopted to minimize the presence of foreign materials. If metal gets into a mill, an overload switch on the mill motor will help prevent hot surfaces and thereby reduce the dust-explosion hazard, as indicated in Sec. 3-4.

Power-operated rotary (star) valves are often used to transfer dust from a collector to a silo, bin, or hopper. Although this is their primary function, they serve an important additional purpose in preventing the spread of explosion pressure from one plant area to another. Nevertheless, burning dust may pass the rotary valve and serve as an ignition source to start an explosion in downstream equipment. Blanketing a silo, bin, or hopper with inert gas, however, can eliminate that hazard. (Enough inert gas must be added to compensate for the oxygen in the air that passes

*Trademarks of the Walter Legge, Co., Inc., New York.

through the rotary valve, if air is used in the upstream conveying system. A vent should be provided for exit of the off-gas from the silo, bin, or hopper.) Moreover, power to the rotary valve can be shut off by the explosion detector if an explosion-suppression system is installed.

Fire may follow a dust explosion; automatic sprinkler protection usually is prudent in areas where combustible dusts are processed. In addition, sprinklers may be desirable inside collectors and storage bins. Sometimes difficulty may be experienced in fighting a fire inside equipment because of the absence of entry ports; the initial design ought to provide connections for introduction of inert gas or other extinguishing agent to help in quenching a fire.

Fabric filters are sometimes provided with conductive wires to prevent the occurrence of static electricity sparks. Generally, it is not likely that a spark to ground from the fabric or collected dust on the filter would have sufficient energy to ignite a dust cloud (see the discussion of static electricity in Sec. 8-2). In this sense conductive bags may not be necessary unless the wires are used to ground the bag cages. Moreover, wires may break and then could become insulated electrical conductors with sparking potential.

Powder Coating

Plastic polymers are being applied increasingly as powder coatings. Powder coating is covered in Data Sheet 7-27, Loss Prevention Data, of the Factory Mutual Engineering Corp. Application of combustible powders is included in *NFPA* 33, Spray Application Using Flammable and Combustible Materials. In addition, the fundamentals of powder coating are covered in Ref. 30.

In spray application, air pressure and electrostatic charging are used to direct powder onto the glass or metallic target by an electrostatic spray gun. Coating thickness applied by spraying is thin, 25 to 400 μm. The article to be coated usually is brought into, and taken out of, a spray booth by a conveyor system. Airflow carries undeposited powder (overspray) down to the bottom of the booth and thence to the powder-recovery system through ducts. The powder-recovery system usually consists of a fabric filter, although a cyclone collector may sometimes be used ahead of the filter. Filtered air may be recirculated back to the building through an absolute filter to save heat.

Several fundamental explosion-prevention and explosion-protection practices that have been described in this chapter and in Chaps. 3 and 5 need to be incorporated into the design and operation of powder-coating systems. Some of these practices for spray application systems are noted below. (Other safety requirements and some of the following practices are included in *NFPA* 33 and Factory Mutual Engineering Corp. Data Sheet 7-27.)

1. Airflow must be sufficient to keep the concentration of powder below the MEC, based on the *maximum* rate of powder discharge from the spray guns. (At the outlet of the spray gun, powder concentrations will exceed the MEC. Thus, the indicated concentrations are averages, determined by dividing airflow by maximum rate of powder discharge from all the spray guns.) Both Factory Mutual Engineering Corp. and NFPA advise that powder concentration not exceed 50 percent of the MEC. If explosibility data are not available, 30 mg/L may be used for the MEC. NFPA requires use of explosion-suppression equipment if the design exhaust-dust concentration is more than 50 percent of the MEC. Moreover, for fixed powder-application equipment whatever the concentration, NFPA requires, also,

> installation of approved flame detection apparatus that will, in event of ignition, react to presence of flame within one-half second and:
>
> **a.** Shut down all energy supplies (electrical and compressed air) to conveyor, ventilation, application, transfer and powder collection equipment.
>
> **b.** Close segregation dampers in associated ductwork to interrupt airflows from application equipment to powder collectors.
>
> **c.** Actuate an alarm.

Airflow can be determined from fan capacity. It is also advisable to measure airflow periodically or continuously to assure its adequacy. Portable techniques to determine airflow in ductwork are given in Refs. 12 and 13. Preventive maintenance is required to prevent leakage of air into ducts. Such leakage could result in much higher concentration of powder upstream of the leak point than might be apparent from airflow estimates or measurements.

2. Even though powder is conveyed below its MEC, propagation of a fire at the gun to the fabric filter is conceivable by burning of powder accumulated in ductwork. Also, if built-up powder is dislodged for any reason, at that moment powder concentrations may be above the MEC. Design and operating procedures should minimize powder buildup. Horizontal surfaces should be minimized to reduce powder accumulation. Housekeeping is important from the safety consideration since any quiescent dust layer can be dislodged or stirred into a dust cloud that could be in the explosive range. Frequent scheduled cleaning can limit dust buildup.

3. All metal parts of the spray booth, workpieces, and powder-collection system must be electrically grounded with resistance to ground not exceeding 1 MΩ. (Hangers are often used to hold objects on a conveyor. Powder may be sprayed onto these hangers and then a hard plastic coating on them can result from curing in the postheat oven. Consequently, metal objects being coated while suspended from the hangers may not be grounded.)

4. Ducts from explosion vents should be directed to the outside of the building through as short a run as possible. (With explosion venting into a building a secondary explosion is possible from powder that has accumulated on beams, ledges, and other structural members. A secondary explosion may also occur in these circumstances if some of the inventory of unburned powder is discharged and then burns in the building in event of an explosion in a vented collector.)

References

1. Vervalin, C. H., "Will Your Plant Burn Tomorrow?," *Hydrocarbon Process.* vol. 54, no. 12, pp. 145–154, December 1975.

2. Yowell, R. L., "Bisphenol-A Dust Explosions," *Chem. Eng. Prog. 2d Loss Prev. Symp., St. Louis, vol. 2, 1968,* pp. 29–33.

3. Jacobson, M., J. Nagy, A. R. Cooper, and F. J. Ball, "Explosibility of Agricultural Dusts," *U.S. Bur. Mines Rep. Invest.* 5753, 1961.

4. Jacobson, M., J. Nagy, and A. R. Cooper, "Explosibility of Dusts Used in the Plastics Industry," *U.S. Bur. Mines Rep. Invest.* 5971, 1962.

5. Nagy, J., H. G. Dorsett, Jr., and A. R. Cooper, "Explosibility of Carbonaceous Dusts," *U.S. Bur. Mines Rep. Invest.* 6597, 1965.

6. Dorsett, H. G., Jr., and J. Nagy, "Dust Explosibility of Chemicals, Drugs, Dyes, and Pesticides," *U.S. Bur. Mines Rep. Invest.* 7132, May 1968.

7. Dorsett, H. G., Jr., M. Jacobson, J. Nagy, and R. P. Williams, "Laboratory Equipment and Test Procedures for Evaluating Explosibility of Dusts," *U.S. Bur. Mines Rep. Invest.* 5624, 1960.

8. Hartmann, I., and J. Nagy, "Inflammability and Explosibility of Powders Used in the Plastics Industry," *U.S. Bur. Mines Rep. Invest.* 3751, 1944.

9. Nagy, J., and D. J. Surincik, "Thermal Phenomena During Ignition of a Heated Dust Dispersion," *U.S. Bur. Mines Rep. Invest.* 6811, 1966.

10. Perry, J. H., *Chemical Engineers' Handbook,* 3d ed., sec. 15, McGraw-Hill New York, 1950.

11. Moodie, T. W., "Measurement Comes to Hazardous Dust Area Classification," *ISA Trans.,* vol. 10, no. 3, pp. 224–230, 1971.

12. Bureau of National Affairs, Inc., EPA Regulations on Standards of Performance for New Stationary Sources *(40 CFR 60), Environ. Rep., Fed. Regul.,* p. 121:-1501.

13. Bloomfield, B. D., chap. 28, "Source Testing," pp. 487–536, in A. C. Stern (ed.), *Air Pollution,* vol. II, Academic, New York, 1968.

14. Liebman, I., R. S. Conti, and K. L. Cashdollar, "Dust Cloud Concentration Probe," *Rev. Sci. Instrum.,* vol. 48, no. 10, pp. 1314–1316, October 1977.

15. Eckhoff, R. K., "Towards Absolute Minimum Ignition Energies for Dust Clouds?," *Combust. Flame,* vol. 24, pp. 53–64, February 1975.

16. Eckhoff, R. K., and G. Enstad, "Why Are 'Long' Electric Sparks More Effective Dust Explosion Initiators than 'Short' Ones?," *Combust. Flame,* vol. 27, August 1976, pp. 129–131.

17. Hay, D. M., and D. H. Napier, "Minimum Ignition Energy of Dust Suspensions," *Inst. Chem. Eng. Symp. Ser.* 49, *Proc. 6th Symp. Chem. Process Hazards Spec. Ref. Plant Des., 1977,* pp. 73–81.

18. Hughes, J. F., and A. W. Bright, "Electrostatic Hazards Associated with Powder Handling in Silo Installations," *IEEE Trans. Ind. Appl.,* vol. IA–15, no. 1, pp. 100–103, January-February 1979.

19. Cooper, W. F., "The Practical Estimation of Electrostatic Hazards," *Brit. J. Appl. Phys.,* vol. 4, suppl. 2, pp. S71–S77, 1953.

20. Kissell, F. N., A. E. Nagel, and M. G. Zabetakis, "Coal Mine Explosions: Seasonal Trends," *Science,* vol. 179, Mar. 2, 1973, pp. 891–892.

21. Palmer, K. N., and P. S. Tonkin, "The Ignition of Dust Layers on a Hot Surface," *Combust. Flame,* vol. 1, March 1957, pp. 14–18.

22. Mitchell, N. D., "New Light on Self-Ignition," *Natl. Fire Prot. Assoc. Q.,* vol. 45, no. 2, pp. 3–10, October 1951.

23. *Inst. Chem. Eng. Loss Prev. Bull.* 008, London, 1975.

24. LeVine, R. Y., "Electrical Safety in Process Plants . . . Classes and Limits of Hazardous Areas," *Chem. Eng.,* vol. 79, no. 9, pp. 51–58, May 1, 1972.

25. Nagy, J., H. G. Dorsett, and M. Jacobson, "Preventing Ignition of Dust Dispersions by Inerting," *U.S. Bur. Mines, Rep. Invest.* 6543, 1964.

26. Hartmann, I., A. R. Cooper, and M. Jacobson, "Recent Studies on the Explosibility of Cornstarch," *U.S. Bur. Mines Rep. Invest.* 4725, August 1950.

27. Palmer, K. N., *Dust Explosions and Fires,* Chapman & Hall, London, 1973.

28. McKinnon, G. P. (ed.), *Fire Protection Handbook,* 14th ed., sec. 2, chap. 2, National Fire Protection Association, Boston, 1976.

29. Nagy, J., A. R. Cooper, and J. M. Stupar, "Pressure Development in Laboratory Dust Explosions," *U.S. Bur. Mines Rep. Invest.* 6561, 1964.

30. Miller, E. P., and D. D. Taft, *Fundamentals of Powder Coating,* Society of Manufacturing Engineers, Dearborn, Mich., 1974.

31. Halm, R. L., "Low Ignition Energies," *Chem. Eng. News,* vol. 57, no. 33, pp. 2, 51, Aug. 13, 1979.

Additional Reference

Gibson, N., "Electrostatic Hazards in Filters," *Filtr. Sep.,* vol. 16, no. 4, pp. 382–386, July-August 1979.

A

Conversion Factors*

A-1 Concentration

Milligrams gas per liter air β at 0°C and 101.325 kPa

$$\times \frac{2.24146}{MW + 0.022415\,\beta} = \%\ v/v$$

Ounces (avoirdupois) per cubic foot air

$$\times\ 1.001154 \times 10^3 = \text{milligrams per liter air}$$

$$\%\ v/v \times \frac{44.6138\ (MW)}{100 - \%\ v/v} = \text{milligrams gas per liter air at } 0°C \text{ and } 101.325 \text{ kPa}$$

A-2 Energy

To convert to joules:

Unit	Multiply by	Unit	Multiply by
Btu_{IT}	1.055056×10^3	$kcal_{IT}$	4.18680×10^3
$ft{\cdot}lb_f$	1.355818	$kg_f{\cdot}m$	9.806650
cal_{IT}	4.18680	L·atm	101.3250

*Following ASTM practice, numbers with zero in the last decimal place are exact, and all subsequent digits are zero. Flow-of-gas figures are carried to six decimal places and are not exact. Otherwise, numbers are rounded to seven significant digits unless such high precision is unwarranted.

IT = International Table
Units IT/thermochemical = 0.999331
Units IT/mean = 1.00077
T = Temperature, K = °C + 273.15

†22.4146 L/mol at 0°C and 101.325 kPa. MW is for gas.

A-3 Flow

To convert to cubic meters per second:

Unit	Multiply by	Unit	Multiply by
ft^3/m	0.4719×10^{-3}	m^3/h	0.2778×10^{-3}
ft^3/s	28.3168×10^{-3}	m^3/min	16.6667×10^{-3}

A-4 Flow of Gas (101.325 kPa)

To convert to cubic meters per second at *T*:

Unit	Multiply by	Unit	Multiply by
kg/h	$23T \times 10^{-6})/MW$	mol/s	$82T \times 10^{-6}$
kg/s	$(82.060T \times 10^{-3})/MW$	lb/h	$(10T \times 10^{-6})/MW$

A-5 Pressure

To convert to kilopascals:

Unit	Multiply by	Unit	Multiply by
atm (standard)	101.3250	mmHg (0°C)	0.1333
bars	100.0	lb_f/ft^2	47.8803×10^{-3}
ftH_2O (4°C)	2.989	lb_f/in^2	6.894757
kg_f/m^2	9.8066×10^{-3}	torr (0°C)	0.1333
millibars	0.10	inHg (0°C)	3.386

A-6 Specific Energy

To convert to megajoules per kilogram:

Unit	Multiply by	Unit	Multiply by
Btu_{IT}/lb	2.3260×10^{-3}	$kcal_{IT}/kg$	4.18680×10^{-3}
cal_{IT}/g	4.18680×10^{-3}		
$kcal_{IT}/mol$	4.18680/MW		

A-7 Velocity

To convert to meters per second:

Unit	Multiply by	Unit	Multiply by
ft/min	5.080×10^{-3}	knots	0.5144444
ft/s	0.30480	mi/h	0.447040
km/h	0.2777778		

A-8 Volume

To convert to cubic meters:

Unit	Multiply by
ft^3	28.3168×10^{-3}
gal (U.S. liquid)	3.7854×10^{-3}

To convert to cubic meters at T and 101.325 kPa:

Unit	Multiply by
kg gas	$(82.06T \times 10^{-3})/MW$
mol gas	$82T \times 10^{-6}$
lb gas	$(37.222T \times 10^{-3})/MW$

B

Equilibrium-Venting Equation

For gas discharge through rupture-disk holders (safety heads) from *Tech. Pap.* 410 of the Crane Company (Ref. 43 of Chap. 5)

$$w = 44.68 Y A_v \sqrt{\frac{\Delta P \; \rho_1}{K}} \qquad \text{kg/s}$$

The subscript 1 indicates upstream conditions, i.e., in the vessel. The factors K and Y can be determined from the procedures outlined in Ref. 43 of Chap. 5.

$$P_1 = p_b + 101 \text{ kPa abs}$$

$$\rho_1 = \frac{(\text{MW})(P_1)}{R_u T_1} \times 10^{-3} \qquad \text{kg/m}^3$$

Thus,

$$w = 15.5 Y A_v \sqrt{\frac{(\Delta P)(P_1)(\text{MW})}{K T_1}} \qquad \text{kg/s}$$

From the equation of state

$$\frac{dP_1}{dt} = r = \frac{w R_u T_1}{(\text{MW})(V)} \qquad \text{MPa/s}$$

For equilibrium venting, dP_1/dt from mass discharge w equals r. Therefore,

$$\frac{A_v}{V} = \frac{7.76 r}{Y} \sqrt{\frac{(K)(\text{MW})}{\Delta P \; P_1 T_1}} \qquad \text{m}^{-1}$$

C

Dispersion Equations

C-1 Dense Stack Gases, Stack Emission

From Eq. (6-1)

$$h_s = 1.58D \left(\frac{v_s}{u}\right)^{0.513} \left(\frac{\chi_s}{\chi_m}\right)^{0.513} \times 10^{-3} - 2H$$

and from Eqs. (6-2) and (6-3)

$$2H = \frac{12.3D^{2/3} v_s A \times 10^{-3}}{u^{1/3}}$$

For specified dilutions of the stack effluent χ_s/χ_m, required h_s is a maximum when $\partial h_s/\partial v_s = 0$. Thus,

$$\frac{\partial h_s}{\partial v_s} = \frac{0.810D(\chi_s/\chi_m)^{0.513} \times 10^{-3}}{u^{0.513} u_s^{\,0.487}} - \frac{12.3D^{0.667}A \times 10^{-3}}{u^{1/3}} = 0$$

Therefore, with $\chi_m = L/5$

$$v_{s,\,\mathrm{crit}} = \frac{3.73D^{0.684}(5\chi_s/L)^{1.05} \times 10^{-3}}{u^{0.368}A^{2.05}}$$

Substitution of $v_{s,\,\mathrm{crit}}$ into Eq. (6-1) yields

$$h_s = \frac{43.8D^{1.35}(5\chi_s/L)^{1.05} \times 10^{-6}}{u^{0.702}A^{1.05}}$$

C-2 Ground-Level Releases E Stability

From Table 6-1

$$\sigma_y \sigma_z = 0.00642x^{1.76}$$
$$\sigma_z^{\,2} = 0.00542x^{1.68}$$

From Eq. (6-6) at the centerline of the plume ($y = 0$)*

$$\chi\ (x,y,z) = \frac{9.916Q \times 10^3}{ux^{1.76}} \exp\left(-\frac{92.25z^2}{x^{1.68}}\right)$$

Therefore, at a given z, concentrations are a maximum at $x_{m\,1}$, *where* $\partial\chi/\partial x = 0$

$$\frac{\partial \chi}{\partial x} = 88.14z^2 - x_{m_1}{}^{1.68} = 0$$

Thus,

$$x_{m_1} = 14.38z^{1.19}$$

With 100 mol/s, for example, the maximum height to which a peak concentration of 1% v/v extends for $u = 1$ m/s is determined as follows:

$$Q = 100 \text{ mol/s} = 2.447 \text{ m}^3\text{/s}$$
$$\sigma_y = 0.0873x^{0.92}$$
$$\sigma_z = 0.0736x^{0.84}$$

From Eq. (6-6)*

$$1 = \frac{(100)(2.447)(2)}{(\pi)(0.00642x_{m_1}{}^{1.76})(1)} \exp\left(-\frac{92.25z^2}{x_{m_1}{}^{1.68}}\right)$$

Since

$$x_{m_1} = 14.38z^{1.19} \text{ m}$$
$$z = 8.01 \text{ m}$$

and

$$x_{m_1} = 171.0 \text{ m}$$

For this case the fictitious upwind source occurs where calculated χ = 100 percent at $y = z = 0$*

$$100 = \frac{(100)\ (2.447)\ (2)}{(\pi)(0.00642x^{1.76})\ (1)}$$
$$x = 22.65 \text{ m}$$

and so

$$x_{m_2} = 171.0 - 22.65 = 148.4 \text{ m}$$

*Momentary concentrations 2χ from Eq. (6-6).

Index